BASIC
STEEL DESIGN

Sears Tower, Chicago, the world's tallest building.

SECOND EDITION

BASIC
STEEL DESIGN

BRUCE G. JOHNSTON
Professor Emeritus, University of Michigan

FUNG-JEN LIN
Principal Engineer, The Ralph M. Parsons Company, Pasadena
Lecturer, University of Southern California, Los Angeles

T. V. GALAMBOS
Professor, Washington University, St. Louis

Prentice-Hall, Inc.
Englewood Cliffs, New Jersey

Library of Congress Cataloging in Publication Data

Johnston, Bruce Gilbert, date
  Basic steel design.

  Includes bibliographical references and index.
  1. Building, Iron and steel. 2. Structural
design. I. Lin, Fung-Jen, joint author. II. Galambos,
Theodore V., joint author. III. Title.
TA684.J6 1980   624′.1821   79-13902
ISBN 0-13-069344-8

Civil Engineering and Engineering Mechanics Series

N.M. NEWMARK AND W.J. HALL, *Editors*

Printed in the United States of America

10 9 8 7 6 5 4 3 2

PRENTICE-HALL INTERNATIONAL, INC., *London*
PRENTICE-HALL OF AUSTRALIA PTY. LIMITED, *Sydney*
PRENTICE-HALL OF CANADA, LTD., *Toronto*
PRENTICE-HALL OF INDIA PRIVATE LIMITED, *New Delhi*
PRENTICE-HALL OF JAPAN, INC., *Tokyo*
PRENTICE-HALL OF SOUTHEAST ASIA PTE. LTD., *Singapore*
WHITEHALL BOOKS LIMITED, *Wellington, New Zealand*

CONTENTS

Chapter 10   Special Topics in Beam Design   303

Chapter 11   Computer-Aided Technology   336

Index   347

FOREWORD

This book, coordinated as it is with the American Institute of Steel Construction Specification for the Design, Fabrication and Erection of Structural Steel for Buildings, must keep fully abreast of changes in that specification if it is to fulfill its purpose. The first edition, published in 1974, was based on the 1969 version of the specification, as modified by supplements in 1970 and 1971. This second edition is based on the new 1978 edition of the specification. Moreover, on September 1, 1978, the rolling mills of the major steel producers in this country initiated production of a new series of wide-flange shapes that has required revision of most of the illustrative examples provided in this second edition.

The authors of the first edition welcome the addition of a new coauthor, T. V. Galambos, Professor in Civil Engineering at Washington University at St. Louis. Dr. Galambos has been project director of research sponsored by the American Iron and Steel Institute to develop criteria for the oncoming load and resistance factor design procedure.

PREFACE

This book is concerned with the basics of structural steel design. It is suitable for reference use or as a text and is unique in at least two respects: (1) reference is made primarily to a single design specification—that of the American Institute of Steel Construction, and (2) the chapters on individual member design include flow diagrams, similar to those used to guide the development of a computer program. These diagrams have been found to be excellent teaching aids.

The use of complex theory is minimized. The purpose of design is to produce a structure. In the design of monumental structures, elaborate analyses, in many cases involving use of the computer, may be required. But in the initial study of steel design, the acquisition of a basic understanding of structural behavior and the meaning of specification requirements can best be attained by simplicity of approach and emphasis on the development of sound structural judgment.

Chapter 1 is a broad and descriptive introduction to the steel structure, covering the properties of steel, the history of the development of steel structures, and touching on the topics of economy, safety, planning, fabrication, construction, and maintenance.

Chapters 2 through 7 are devoted to the various types of structural members in common use: the tension member, the beam, the column, and so on. Each of these chapters takes up the structural behavior problem, explains pertinent AISC Specification clauses, and summarizes (with the exception of Chapter 6) the logical application of the specification by means of flow diagrams. Although the flow diagram was developed primarily as an aid to the development of a computer program, it also serves admirably as a summary and as a guide to the logical sequence of steps that must be taken in the design selection of a particular structural member.

Chapter 8 goes beyond the treatment of the individual member to provide a study of both the elastic and plastic design of continuous beams and frames. It provides a review and summary of the moment-distribution method as applied to the design of such structures.

Chapter 9, on load and resistance factor design, like Chapter 8, is entirely new. It represents a probable future trend that introduces probabilistic and statistical procedures to achieve a better and more rational balance between economy and safety in design. This material was prepared by T. V. Galambos, director of research on this subject as sponsored at Washington University by the American Iron and Steel Institute.

Chapter 10 covers special topics that are of occasional importance, but less common than those treated in the earlier chapters. These topics include the torsion of both open and box members, combined bending and torsion, biaxial bending, lateral-torsional buckling, and the shear center.

A final new addition, Chapter 11, briefly summarizes various aspects of computer-aided design.

Steel design specifications are essentially similar, but the diversity of formulas pertaining to identical problems (such as the column design formulas) is confusing to the beginning student. But after learning to design with a particular specification the student can readily adapt his basic knowledge to a different one.

The purchase of the *Steel Construction Manual* of the American Institute of Steel Construction is essential to the complete use and understanding of this book. The AISC *Manual* also includes the AISC Specification, which will be referred to throughout the book. Moreover, practice in use of the AISC *Manual* is essential as a secondary educational objective toward structural design practice. The nomenclature that will be used herein is nearly identical with that found in the AISC Specification and will not be repeated herein, except in individual references to the presentation of equations or formulas.

The flowchart symbols used in Chapters 2 through 9 have the following significance:

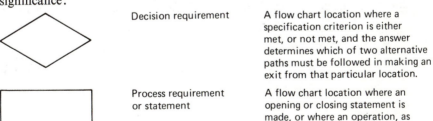

| | Decision requirement | A flow chart location where a specification criterion is either met, or not met, and the answer determines which of two alternative paths must be followed in making an exit from that particular location. |
| | Process requirement or statement | A flow chart location where an opening or closing statement is made, or where an operation, as stated, is to be performed. |

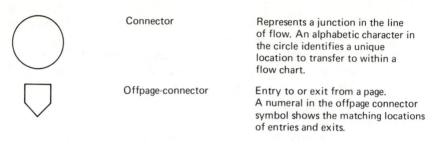

| Connector | Represents a junction in the line of flow. An alphabetic character in the circle identifies a unique location to transfer to within a flow chart. |
| Offpage-connector | Entry to or exit from a page. A numeral in the offpage connector symbol shows the matching locations of entries and exits. |

Special acknowledgement is made to William Milek, Jr., of the American Institute of Steel Construction, for his cooperation in the preparation of this book, and to Frank W. Stockwell, Jr., also of AISC, for his detailed check of much of the final draft of the first edition. Thanks are also due to Randolph F. Thomas for his help in the preparation of many of the drawings. The first author has appreciated the support of his wife, Ruth, and will always be indebted to his father, the late Sterling Johnston, for having inculcated from an early age an empathy with steel structures. The second author has always appreciated the support of his parents, Huan-Yu and Chang, and the University of Michigan, where he completed his doctoral studies. The Prentice-Hall staff, through Christopher Moffa and Lori Opre, have been most cooperative in expediting the processing of the manuscript. Acknowledgment is also due to Sterling Johnston, in charge of computer services for structural analysis and design at TVA, for his review and suggested revisions of Chapter 11.

Each chapter includes a number of example problems which are presented with more complete details than would be required by an experienced designer. Problems for assignment are also included, with initial emphasis on variations of the example problems, thus adding incentive to the careful study of the examples in the text.

A word about SI units and the trend toward their adoption. It would be counter to the essential purpose of this book in its 1 : 1 relationship with the AISC Specification and Manual not to use the same units. All specification formulas and tables would need to be rewritten and the result would be confusing to the reader. In any future edition of this book we will, of course, adopt the same units as are then used by the related specification and manual.

BRUCE G. JOHNSTON
FUNG-JEN LIN
T. V. GALAMBOS

ABBREVIATIONS

AISC	American Institute of Steel Construction
AISCM	AISC Manual of Steel Construction
AISCS	AISC Specification for the Design, Fabrication and Erection of Structural Steel for Buildings, Nov. 1, 1978 edition.
AISI	American Iron and Steel Institute
ASCE	American Society of Civil Engineers
ASD	Allowable stress design
ASTM	American Society for Testing and Materials
C	Channel shape
CAT	Computer aided technology
CRC	Column Research Council
FLB	Limit state moment for flange local buckling
FS	Factor of safety
L	Angle shape
LF	Load factor
LRFD	Load and resistance factor design
LTB	Limit state moment for torsional buckling
PD	Plastic design
PL	Plate
S	Standard beam shape
SF	Shape factor
SSRC	Structural Stability Research Council
ST	Structural tee cut from S shape
TS	Structural tubing
W	Wide-flange shape
WLB	Limit state moment for web local buckling
WT	Structural tee cut from W shape
WW	Wide-flange (W) shape made by welding 3 plates

1

THE STEEL STRUCTURE

1.1 INTRODUCTION

It is only by means of *structure* that the observable external details of our planet's surface are altered. Structures are the earmarks of our civilization, and the structural engineer—through the practice of construction within the framework of civil engineering—helps to create them: the buildings, dams, bridges, power plants, and towers that make possible our shelter, power, transportation, and communication. Thus the civil engineer has a responsible role in determining whether or not the structures that he builds enhance or detract from their environment.

After the prospective owner of a structure has considered alternatives and selected the site and has made subsurface exploration of soil conditions, the structural design is initiated by a consideration of various structural systems, alternative types and disposition of members, and the preparation of preliminary design drawings. Subsequently, the structural designer determines the required sizes of members and their connections, describing these in detail through drawings and written notes, so as to facilitate the fabrication and construction of the structural frame. One must first learn to design the parts before he can plan the whole. Hence the emphasis herein is on the design and selection of steel tension members, beams, compression members (columns), beam-columns, plate girders, and connections that join these members to form a bridge, building, tower, or other steel structure. In addition, attention is given to the design of simple frames that involve an assemblage of members into a structure.

The adequacy of a structural member is in part determined by a set of design rules, called specifications, which include formulas that guide the designer in checking the strength, stiffness, proportions, and other criteria that may govern the acceptability of the member. There are a variety of specifications that have been developed for both materials and structures. Each is based on years of prior experience gained through actual structural usage. The diversity of specification formulas and rules pertaining to essentially similar problems is a source of confusion in the study of structural design. In this book reference will be made primarily to a single specification —the widely used American Institute of Steel Construction (AISC) *Specifications for the Design, Fabrication and Erection of Structural Steel for Buildings*, as revised in 1978. He who masters the use of this specification, and understands the structural meaning and significance of its requirements, can readily turn to some other specification pertaining to the design of steel structures and understand the parallel set of design rules that it will contain.

The 1978 AISC Specification will be found in the eighth (1979) edition of the AISC *Manual of Steel Construction* along with much additional design information and tabular data. This AISC Manual must be considered to be an essential companion volume to this book and frequent reference will be made to it. To abbreviate the repeated references that will be made to the manual and the specification, they will be referred to herein as AISCM and AISCS, respectively.

At this point one should read the Foreword and Preface in the AISCM and thumb briefly through the entire book to get a preliminary idea of its contents.

1.2 THE STRUCTURE AND ITS PARTS

The basic framework that gives strength and form to a structure does so in the same way that the human skeleton gives strength and form to the human body, and in accordance with the same principles.

The creation of the complete structure calls for the combined services of the architect, civil engineer, environmentalist, city planner, and other specialists in engineering fields that may include acoustics, machine design, illumination, heat, ventilation, and other facilities. The overall design, processing, and scheduling of these inputs and the consideration of the involved interrelations while a structure is being planned and built have become known as *systems engineering*.

A steel design text must be concerned initially with the structural members that are the component parts of the overall structure. In a steel structure these are beams, which carry loads transverse to their long axis, columns or compression members, which transmit compression force along their long axis (the trunk of a tree is a most efficient column), and tension members, exemplified by a wire rope, most effectively capable of transmitting tension force, or pull, and built up of many individual wires that have been cold drawn to greatly increase their strength. Compression members usually must also carry some transverse loads and as such are called *beam-columns*. The way a structure is made up of these component parts is illustrated in Fig. 1.1,

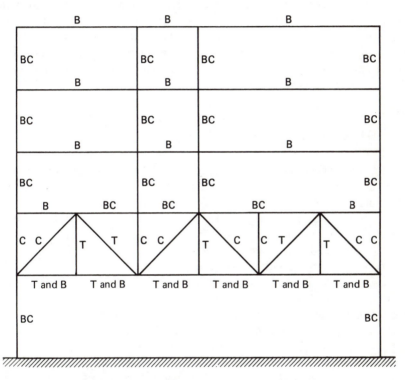

Fig. 1.1 Structural frames are composed of beams (B), columns (C), tension members (T), and beam-columns (BC).

in which the upper part of a building frame is carried over an auditorium by means of a truss. In this figure the columns, beams, beam-columns, and tension members are labeled by the letters C, B, BC, and T, respectively. At every juncture point or joint between the ends of the members, connections must be provided, often offering the most challenging design problems, as they are the least standardized—yet essential to the continuity of the structure and its resistance to collapse.

A good structural designer *thinks* about the actual structure as much or more than the mathematical model that he uses to check the internal forces for which he will determine the required material and type, size, and location of the members that carry the loads. The "structural engineering mind" is one that can visualize the real structure, the loads upon it, and in a sense "feel" how these loads are transmitted through the various members down into the foundations. The best designers are gifted with what sometimes has been called "structural intuition." To develop "intuition" and "feel," the engineer is a keen observer of other structures. He may even contemplate the behavior of a tree, designed by nature to withstand violent storms, flexible where it is frail in leaves and small branches, but growing in strength and never abandoning continuity as the branches merge with the trunk, which in turn spreads below its base into the root system, which provides its foundation and connection with the earth.

1.3   STRUCTURAL STEEL†

A knowledge of the elastic, inelastic, fracture, and fatigue characteristics of a metal is needed for the evaluation of its suitability for the making of a structural member for a particular structural application. *Elasticity* is the ability of a metal to return to its original shape after loading and subsequent unloading. *Fatigue* of a metal occurs when it is repeatedly stressed above its *endurance limit* through many cycles of loading and unloading. *Ductility* is the ability to be deformed without fracture in the inelastic range—that is, beyond the *elastic limit*. In steel, loaded in a simple tension state of stress, there occurs a sharp *yield point* at a stress only slightly greater than the elastic limit. When loaded beyond the yield point, the ductility of structural steel permits it to experience large *inelastic* elongation. Finally, the ultimate breaking strength is reached and the specimen fractures. The tensile load at fracture, divided by the original area of the unloaded specimen, is termed the *ultimate tensile strength*. Minimum specification values of the yield point, ultimate tensile strength, ductility indices, and chemistry have been established by the American Society for Testing and Materials (ASTM) to control the acceptance of structural steels.

Section 1.4.1.1 of the AISCS lists 14 steels that are used in the manufacture of structural steel products. Of these, those listed in Table 1.1 are used in structural steel shapes and plates and will be designated in design examples and problems to follow.

†The reader may wish to supplement the study of this section by reference to Chapter 1, "The Structural Steels and Their Mechanical Properties," of Ref. 1.7.

Table 1.1

Steels Used in Structural Steel Shapes and Plates

Designation	Minimum yield point (ksi)†	Ultimate strength (ksi)†
Structural steel, ASTM A36	36	58–80
Hot-formed welded and seamless carbon steel structural tubing, ASTM A501	36	58 min.
High-strength low-alloy structural steel, ASTM 242		
High-strength low-alloy structural manganese–vanadium steel, ASTM 441	⎧40‡ ⎪42 ⎨46 ⎩50	60⎫ 63⎪min. 67⎬ 70⎭
High-strength low-alloy structural steel with 50,000 psi minimum yield point to 4 in. thick, ASTM A588		
Hot-formed welded and seamless high-strength low-alloy structural tubing, ASTM A618	⎧50 ⎩50	65⎫min. 70⎭
High-strength low-alloy columbium–vanadium steels of structural quality, ASTM A572	⎧42 ⎪50 ⎨60 ⎩65	60⎫ 65⎪min. 75⎬ 80⎭
High-yield-strength quenched and tempered alloy steel plate, suitable for welding, ASTM A514	⎧ 90 ⎩100	100–130 110–130

†ksi stands for "kips per square inch"; 1 kip = 1000 lbs. Although there is a general trend to convert to metric units, no recognition has as yet been given to this move in the 1978 AISC Specifications that are the basis for this book, and it is therefore impracticable to introduce metric notation in this edition.

‡ASTM 441 only.

The mechanical properties of structural steel that describe its strength, ductility, and so forth, are given in terms of the behavior in a simple tension test. The initial portion of a typical tension stress–strain curve for structural steel is shown in Fig. 1.2(b). To a greatly different horizontal scale, the complete curve is given in Fig. 1.2(a) but the load-carrying capacity of beams and columns is very largely determined within the range of Fig. 1.2(b). The slope of the stress–strain curve in the elastic range is termed E, the modulus of elasticity, and is taken as 29,000 kips per square inch (ksi) for the structural steels. The yield point, F_y, is the most significant property that differentiates the structural steels.

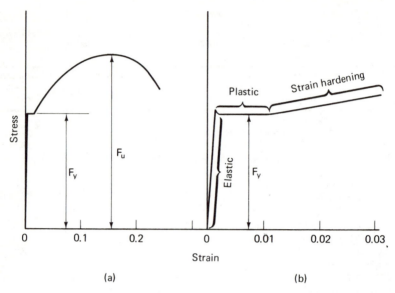

Fig. 1.2 Typical stress–strain curves from a tension test of structural steel.

The yield point of steel will vary somewhat with temperature, speed of test, and the characteristics (size, shape, and surface finish) of the test specimen. After initial yield, the specimen elongates in the *plastic* range without appreciable change in stress. Actually, yield occurs at very localized regions, which *strain-harden*, that is, strengthen, so as to force yielding into a new location. After all the elastic regions have been exhausted, at strains of from 4 to 10 times the elastic strain, the stress starts to increase and a more general strain hardening or strengthening commences. The sharp yield point and flat yield stress level shown in Fig. 1.2(a) are peculiar to the non-heat-treated structural steels.

Structural steels are unique in that they are tough. *Toughness* may be defined as a combination of strength and ductility. After the general strain-hardening range commences in the tension test, the stress continues to increase, and inelastic extension of the test specimen continues uniformly (without local reduction in cross-sectional area) until the maximum load is reached. The specimen then experiences a local constriction and is said to *neck down*. The nominal stress based on the original area is termed the *ultimate tensile strength*, F_u, of the material. The ability of steel to withstand inelastic deformation without fracture also permits it to sustain local yielding during fabrication and construction, thus allowing it to be sheared, punched, bent, and hammered without apparent damage.

Under certain combinations of circumstances, steel structures may develop cracks without appreciable prior ductile deformation. The designer should avoid sharp reentry corners that cause stress concentrations, especially in large boxlike or tank structures. Sheared edges and punched holes also cause minute stress concentrations and damage the edge material where cracks are apt to start. Operation in extremely low temperature is another factor conducive to brittle fracture. Thus careful attention to smooth edge preparation, avoidance of stress raisers, and quality control of material and fabrication processes will reduce the likelihood of brittle fracture to a minimum.

Most frequently the designer will use a standard steel shape as a structural member. These are hot rolled from billets, and their standardized dimensions for detailing and properties for design are completely tabulated in the AISCM. These include wide-flange and miscellaneous beams, channels, and angles. The beams range from a depth of 3 inches (in.) up to 36 in. at 300 pounds per foot (lb/ft), and include (primarily for use in tall buildings) a series of very wide column sections having a nominal depth of 14 in. with weights of 90 to 730 lb/ft.

The reader should turn to the AISCM, where the availability and selection of the appropriate grade of steel as well as the availability of shapes, plates, and bars are discussed and tabulated. The abbreviations and symbols used in designating hot-rolled structural steel shapes also are listed in the AISCM and will be used throughout the text and in assigned problems.

In some parts of the country, less accessible than others to those mills that roll heavy shapes, equivalent shapes are made up by welding three plates together. Of course, when either beam or column section requirements exceed those available in standard rolled shapes, such sections are tailor made, as it were, by welding together plates to form heavier girder or column sections.

In addition to hot-rolled shapes, standard sizes of plate, bar, pipe, and hot-formed tubing, either square, rectangular, or circular in cross section, are available.

Supplementing the available range of hot-rolled sections is a wide variety of both standard and special cold-formed shapes. Their use in design is covered by the *Specification for the Design of Cold-Formed Steel Structural Members* (Ref. 1.4) of the American Iron and Steel Institute, which is coordinated as much as possible with the AISCS.

Quoting from the commentary (Ref. 1.5) on the cold-formed member specification,

> members are cold-formed, in rolls or brakes, from flat steel, generally not thicker than $\frac{1}{2}$ and as thin as about 0.0149 in.
>
> Cold-formed members, as distinct from heavier, hot-rolled sections, are used essentially in three situations: (1) where moderate loads and spans render the thicker, hot-rolled shapes uneconomical, (2) where, regardless of thickness, members are wanted of cross-sectional configurations which

cannot economically be produced by hot-rolling or by welding of flat plates, and (3) where it is desired that load-carrying members also provide useful surfaces, such as in floor and wall panels, roof decks, and the like.

The 1977 *Cold-Formed Steel Design Manual* (Ref. 1.8) includes Refs. 1.4 and 1.5 together with "Supplementary Information," "Illustrative Examples," and "Charts and Tables."

1.4 LOADS ON STRUCTURES

The structural designer is required to determine the member shape, size, and arrangement that will safely carry the loads that the structure is expected to experience. The weight of the structure, which must be estimated in advance, together with all permanently attached equipment, is termed the *dead load*. Especially in very tall buildings or long-span structures it is important to check the final design with regard to the adequacy of the initial dead-load estimates. *Live load* consists of stored material, people, vehicles, snow, ice, wind, explosive blast, water in motion, earth pressure, impact, earthquake effects, and so on,—all dependent on the type of structure, its intended use, and its geographic location. Loads can occur in combination, and the probability of such combinations as well as the magnitude of the loads must be considered. Live loads on standard-type structures are generally specified by the various building codes. Structures of unusual shape may require wind-tunnel-model tests to determine the magnitude and distribution of the load. Similarly, loads induced by earthquake may be based on building code requirements for conventional structures but may require dynamic analyses and/or dynamic model tests in the case of unusual structures.

1.5 HISTORICAL DEVELOPMENT

Structural design in ancient times was simply a matter of repeating what had been done in the past, with little knowledge of material behavior or structural theory. Success or failure was determined simply on the basis of whether the building or bridge supported the actual load or collapsed under it. Experience then was the only teacher; it is still today a most important element of good design. Gradually, through centuries of experience, the art of proportioning members evolved. Empirical rules were established. The columns in Grecian temples were said to be proportioned with the slenderness ratio of a woman's leg. The great builders of the Renaissance had no knowledge of stress analyses, yet achieved structures that required more than empiricism. They were artist, architect, engineer, and builder combined, and their cathedral domes stand now as evidence that they were able to intuitively design magnificent structures that today would not be attempted without the use of sophisticated procedures based on mathematical analyses.

Structures of the past and present, and predictions regarding structures of the future, are directly conditioned by the development and commercial availability of structural engineering materials. Certain of these materials, such as stone, brick, timber, and rope, have been used since the beginning of recorded history. Columns of stone blocks, hewn with precision, are dominant features of Egyptian, Greek, and Roman temples. The aqueducts and bridges of Rome were stone arches, which, like columns, transmit primarily compressive stresses. The Stone Age of structures persisted into the early part of the nineteenth century, when most arches and domes were still built of stone masonry and held in place by stone buttresses.

The commercial development of iron provided the first of the structural metals that were to open up an entirely new world to the structural engineer. The first bridge to be constructed completely of cast iron in 1779 still stands at Coalbrookdale in England. But (in bridges) the use of cast iron, which failed with a brittle fracture in tension, was short-lived. The commercial production of wrought-iron shapes in 1783 brought rapid changes, as it made available a product with that added quality of toughness exemplified by an ability to take large tensile deformation in the inelastic range without fracture. Moreover, wrought iron could be formed into flat plates that could be bent and joined by rivets, making possible the steam locomotive, which, in turn, created a demand for long-span metal bridges. Noteworthy among the early wrought-iron bridges was the Britannia Bridge across the Menai Straits of the Irish Sea. It consists of twin parallel box girders continuous over four spans, with two center spans of 460 ft each, flanked by 230-ft end spans. It was completed in 1850 and is the prototype of a current trend in bridge construction that may be called "the rebirth of the box girder bridge."

The development of the Bessemer converter in 1856 and the open-hearth furnace in 1867 introduced structural steel, and this is the material that has been used most in bridges, as well as in many buildings, for the past 100 years. The first major bridge to be constructed entirely of structural steel was the famous Eads Bridge across the Mississippi at St. Louis. Completed in 1874, it incorporates tubular steel arches with a central span of 520 ft, flanked by 502-ft side spans.

Paralleling the development of iron and steel as engineering materials were advances in material-testing techniques and in structural analysis that made possible the transition of structural design from an art to an applied science. Hooke (1660) demonstrated that load and deformation were proportional, and Bernoulli (1705) introduced the concept that the resistance of a beam in bending is proportional to the curvature of the beam. Bernoulli passed this concept on to Euler, who in 1744 determined the elastic curve of a slender column under compressive load. Important developments in the late 1800s included (1) the manufacture of mechanical strain-measuring instruments that made possible the determination of the elastic moduli that related stress to strain, (2) correct theories for the analysis of stress and deformation resulting from either the bending or twisting of a structural member, and (3) the extension of column-buckling theory to the buckling of plates and the lateral-torsional buckling of beams.

The foregoing advances made possible the development of engineering specifications built around the *allowable-stress method* of selecting structural members. The first general specification for steel railway bridges was developed in 1905, and the first highway bridge specification in 1931. In 1923 the AISC brought out its first general specification for building construction. Under each of these specifications, the criterion for acceptable design strength is as follows: the calculated maximum stress, assuming elastic behavior up to anticipated maximum loads, is kept lower than a specified allowable stress. The allowable stress is intended to be less than the calculated stress at failure by a *factor of safety*. Unfortunately, the calculated maximum elastic stress at failure load varies widely. A slender column or laterally unsupported beam may fail at a fraction of the yield point stress, but a very short column will reach the yield point before it fails. A statically loaded tension member may develop the ultimate tensile strength of the material, or nearly twice the yield point; but the same member, loaded and unloaded repetitively for thousands of cycles, may fail due to fatigue at a fraction of the yield point. A connection, because it yields locally, may not fail until the calculated *elastic* stress is several times the yield point; but it, too, is susceptible to fatigue failure at much lower stresses. It is evident that the true criterion of acceptability is strength—not stress—and thus, on the basis of experience and strength analyses, specified allowable stresses have had to be adjusted upward and downward over a wide range to provide a reasonably uniform index of structural strength.

During the past 60 years, and especially during the past 30, increasing attention has been given to the evaluation of the inelastic properties of materials and to the direct calculation of the ultimate strength of a member. This information is useful in improving the allowable-stress procedure, but it also permits bypassing stress calculation by using the calculated member strength as a direct basis for design. *Load-factor* design has resulted. The maximum anticipated service loads are multiplied by a load factor to yield a required strength, which must be less than the directly calculated strength. Philosophically, this is a more realistic, direct, and natural procedure. The load-factor approach has been used for many years in aircraft design, and Part 2 of the AISCS, introduced in 1961, permits it now as an acceptable alternative to allowable-stress procedures for the design of continuous frames in building structures. Although the current trend in design is to deemphasize the calculation of stress, such calculations are still essential in the design of machine parts and structural elements that must endure many load repetitions. Total resultant stresses must also be calculated in truss analysis and design.

Structural design methods will be undergoing rapid change in the decade of the 1980s as more and more specifications put increasing emphasis on load-factor design (also called *load and resistance factor design* in the United States and *limit states design* in Canada). These newer methods will use multiple load factors with the different types of loads (i.e., different load factors for dead, live, wind, and snow loads) and resistance factors by which the computed strength of beams, columns, connectors, and so on, are multipled to account for the various uncertainties inherent in predicting loads and strengths. Furthermore, the load and resistance factors will be determined by probabilistic methods from the statistical data on loads and strengths. Chapter 9 has been included to give the reader a preview of what is to come.

1.6   STRUCTURAL DESIGN ECONOMY

In a competitive world, with increasing costs for materials and labor, the search for greatest design economy consistent with safety and the desired life of the structure is of major importance. Members must be shaped, arranged, and connected in ways that will provide an efficient and economical solution to the design problem, having in mind not only the cost per pound of the material itself, but also the labor costs of shop fabrication and field erection. Minimum weight is often a design goal. However, if simplicity of fabrication is sacrificed to achieve minimum weight, the overall cost may be increased. In Fig. 1.3(a) a steel beam under uniform load will be adequate in strength if it is fabricated as shown out of three segments—two end pieces,

Fig. 1.3 Reducing weight may increase cost.

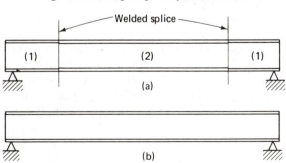

labeled (1), which weigh less per foot of length than the center section, labeled (2). But the cost of welding the three segments together may (or may not) exceed the cost of the added weight if the beam is made of a single nonwelded member having the same size as the center segment, as shown in Fig. 1.3(b).

A similar situation may occur in plate girder design.† The use of very thin webs is made possible by vertical and (in some cases) horizontal stiffeners welded to the web. In borderline instances, the use of a thicker web, which eliminates the need for stiffeners, can result in a saving in fabrication cost even though the overall girder weight is increased.

Figure 1.4 illustrates how member arrangement can affect economy. In

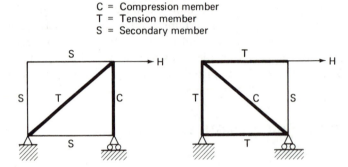

Fig. 1.4 Economy as affected by member arrangement.

each trussed rectangular frame, load H is applied horizontally at the top, acting as shown, and only at A is there horizontal reaction support. In arrangement (a), only two of the five members are directly stressed by the load. These are called *load-bearing* members and are indicated by the heavily weighted lines. But in arrangement (b), four of the five members are stressed by the load. The unstressed members would be called *secondary* members. Moreover, in arrangement (b) the compression (*C*) member is longer than in arrangement (a), thus using more material because of the lesser efficiency of compression members in comparison with tension (*T*) members. The foregoing illustrates the general principle that greatest economy results by providing the most direct path possible for the transmission of force from load point to footing.

Method of shipment can have an important bearing on economy. Connections can be made in the fabricating shop at a fraction of the cost for the same connections made during field erection. A fabricating plant built on a navigable waterway has a great advantage in building a bridge over a river accessible to the same waterway. Girders several hundred feet in length can

†See Chapter 7.

be shop-fabricated with no field splices and shipped direct to the site on barges. The same girders, if shipped by rail or truck, would need several field-splice connections, and if their overall height exceeded rail or highway clearance limitations, horizontal field splices would also be necessary. In the case of major bridges, a temporary fabrication shop may be built adjacent to the site to avoid shipment of bridge segments.

In short-span structures, dead weight contributes but little to the stress. But as the span increases, so does the proportion of the dead-load stress to the total combined stress. Finally, when the span is so great that most of the stress is due to dead load, the upper limit of span has been reached for that material and that type of structure. Thus in long-span bridges or in tall buildings, careful attention to weight reduction and accuracy in dead-weight computations take on increasing importance. In such structures the use of high-strength steels for load-carrying members and lightweight metals for non-load-carrying elements is advantageous.

1.7 STRUCTURAL SAFETY

Structural safety can be assured by a combination of good design, good workmanship in fabrication, and good construction methods. The avoidance of any possibility of structural failure should be a primary concern of the designer.

In design, the choice of a proper load factor for plastic design or proper unit stresses and procedures of analysis for allowable-stress design requires experience and sound engineering judgment. Questions of deterioration due to corrosion during the planned life of the structure, variation in material properties, and many other factors also need to be considered. The most rational approach to the problem of structural safety requires a statistical evaluation of the random nature of all the variables that determine the strength of the structure, on the one hand, and those that may cause it to fail (primarily, the loads) on the other hand. Then, by elementary probability theory, the risk of failure may be evaluated and the probability of its occurrence kept at an acceptable level, dependent on the importance of the structure, risk to human life, and other factors. Increasing attention is being given to this approach to safety evaluation, and statistical studies are being made of material properties, variation in strength of various types of members, and loads. Particular attention also has to be given to uncertain loads, such as those due to wind and earthquake.

Years of design experience, conditioned by both unsuccessful and successful behavior, have produced criteria that aid the choice of safe stress levels. These may not always produce the most economical structure; nevertheless, the overall cumulative experience in engineering design has provided a background from which the engineer derives confidence in many particular design applications. Obviously, great skill, care, and more detailed stress analyses, possibly supplemented by laboratory tests of models or portions of a prototype structure, are needed when the designer attempts a new and venturesome type of structure.

1.8 PLANNING AND SITE EXPLORATION FOR THE SPECIFIC STRUCTURE

After the decision is made to build a steel structure to fill particular service functions, consideration is given to those factors that may influence the overall economy. If alternative sites are available, in the case of large and heavy structures, preliminary explorations are required of the various locations with a mapping of the terrain and partial preliminary study of subsurface foundation conditions by means of borings and/or open-pit excavations. Load-bearing tests may be required. If the terrain is uneven, certain functions of a building may make advantageous use of the changing elevation of the ground, and this will of course affect the overall structural layout. Other influencing factors are transportation facilities, availability of water, gas, and other utilities, drainage features, orientation with respect to prevalent winds, consideration for daylight lighting, and the general type of foundation to be required. The minimization of energy requirements is of steadily increasing importance. After all these considerations and after selection of the exact site, more test borings should be made if there is any question at all about the foundation conditions or their uniformity. It may very well happen that the preliminary borings have straddled some subterranean stream location, hard strata lenses, or rock faults with associated local poor foundation conditions. In such a case, misleading estimates resulting in costly changes in design may be avoided by complete subsurface exploration. In recent years subsurface seismic surveys have proved remarkably accurate in locating bedrock and other hard layers. Such surveys are much cheaper than borings and may be used as a preliminary step covering wide areas, followed up by dry sample borings covering smaller areas selected by the results of the seismic survey. In regions subject to settlement of foundation or questionable soil capacity, undisturbed soil samples should be taken for laboratory tests of unconfined and confined compressive strength, shear strength, degree of consolidation, permeability, and so forth. If pile foundations are used, test piles may be required.

1.9 STRUCTURAL LAYOUT, DETAILS, AND DRAWINGS

After preliminary drawings have been made of the space and area requirements, in plan and in elevation, and general decisions made as to materials, type of structure, and so forth, the designer may proceed with a preliminary trial location of columns and footings.

With respect to both design and fabrication, *economy results from simplicity and duplication.* Duplication also leads to the use of standardized mass-produced elements, such as windows and doors, and has led to what is termed *modular construction.* The module is a basic space dimension that repeats itself throughout the structure and may apply to the column spacing or to smaller details. The module in building construction is frequently about 5 ft, usually a multiple of 4 in. This module is used throughout the building and applies to partitions, ceilings, lighting, windows, and so on. Columns may be four or five modules on center, as part of the modular scheme. Thus standardized spacing results in an increase in duplication and standardization of details. Duplication in floor and roof construction will also result from the constancy of column spacing chosen in the basic modular design concept. Such duplication leads in the fabrication shop to fewer different sizes and lengths of beams in the overall steel order, and the duplication of beam and column details reduces the number of design detail drawings that are required. Repetition speeds up the work in the shop with corresponding reduction in cost.

The choice of roof and exterior wall construction involves the possibility of selecting some commercially developed standardized roof or wall product that will determine within reasonable limits the spacing of purlins and girts. However, within the range of feasible variations in purlin and girt spacing consistent with the modular layout, preliminary designs and cost estimates of various spacings should be made to determine the least weight of steel, and this will usually be of least cost as well.

Procedures for the design of main members, such as beams, columns, and tension members, are fairly simple and precisely laid down by specifications. It is in the design of the connecting details between members and their supports that the structural engineer is called upon for the greatest judgment and design skill. Poorly designed connections may lead indirectly to the failure of main members or even the entire structure. In any structure the load must be transmitted through successive connections from points of application down to the footings. The designer must follow the same sequence—for each succeeding component part of a structure must carry the accumulated dead weight of tributary components, and in preliminary design studies these weights can only be approximated.

Important in the design of details is the elimination of bending or eccentricity in local elements. As one specific example, if a column is supported by a beam (or acts as a local beam support), the webs of the column and beam should be in alignment; but, since the major load in the column is carried in the flanges, bearing stiffeners may be needed to transfer the load from the column flanges to the beam web. Thus the load is transmitted from point to point throughout the structure in the most efficient manner without possibility of local failure.

Careful attention must be given in the design of structural details to the method of fabrication and erection, with proper clearances for bolting or riveting, welding electrodes and holders, or whatever else may be required for the fabrication process in question. The designer of details must visualize the complete construction operation. Care must be taken to provide drainage holes in pockets of exposed steel construction, because moisture and dirt should not be permitted to collect at these points of greatest incipient corrosion.

Design drawing should be complete and easy to follow. The AISC textbook *Structural Steel Detailing* provides an authoritative guide. As in many other separate aspects making up the whole of an engineering design, the *saving* of money in preparation of design drawings may result in greatly *increasing* the overall costs of the structure. If the structure is to be bid upon by private organizations, lack of sufficient detail in the design drawings may cause the bidder to add an appreciable amount for the contingencies that he must be prepared to face later when the details of connections and other framing problems are fully brought to light. If these are completely shown in the initial design drawings, the bidder will be able to give the best possible price.

1.10 FABRICATION METHODS

For greatest economy the design must be made in the light of a preliminary decision as to material and mode of fabrication. For example, the economy due to welding largely results from the introduction of continuity (in either allowable stress or plastic design) and from the elimination of the connecting pieces that would be needed in riveted or bolted construction. Although it is usually uneconomical to use both bolting and welding in the shop fabrication of a particular member, because of the double handling, the designer should consider the use of shop welding with bolted field connections. This is especially appropriate in the case of truss bridges, with high-strength bolts used for the field connections. Such connections have an excellent record for resistance to repeated load.

The use of welding requires careful and competent inspection both with regard to procedure and finished product. Both shop and field inspection of welding are important, as the quality of welds depends to a large extent on the skill, character, and endurance of the welder. Standards for quality of materials, procedures, and inspection of welds and welding processes as established by the American Welding Society are quite generally accepted by the AISCS.

When bolting or riveting is to be used, the question arises as to whether punching of holes, subpunching with reaming, or drilling should be employed. Punching with automatic spacing equipment and repetition of members having the same punch pattern is a very economical means of preparation for bolts and rivets. However, punching damages the material locally at the edges of the holes, and such members are not as good under repeated load as are members with drilled holes. Of course, only such members as will receive large fluctuations in applied load require consideration of fatigue strength. There would be no point in subpunching and reaming—or drilling—holes for the connection of roof purlins to their truss supports, because the maximum loads are repeated a relatively few times and the stresses are minimal. In the case of shop assemblies joining several different plates or members, economy may be achieved by clamping the pieces into a single "pack" for single or multiple drilling through all pieces in one operation. Drilling provides smooth edges of holes and the best possible resistance to repeated load.

1.11 CONSTRUCTION METHODS

Structural designs should be prepared with ample consideration of the manner and facility with which field erection can proceed. The arrangement, number, type, and location of field splices and connections should be planned so as to avoid unnecessary duplication of construction equipment and provide the simplest possible erection plan with a minimum of field work. Connections should be arranged to facilitate field assembly. Careful design planning in relation to construction will minimize the total cost of the project. In important large projects a definite erection plan should be presented, but the contractor should have freedom to exercise his own ingenuity through alternative schemes that meet the approval of the owner.

In one particular sense, proper construction methods have a special relation to overall economy, since it is during construction that failures in engineering structures most often occur. During lifting operations, members of trusses that normally are in tension, or the lower flange of a plate girder, which is normally in tension, may be placed in compression with consequent possible buckling failures. In the case of very long plate girders used in bridge construction, special horizontal temporary truss systems fixed to the plate girders may be used during erection. Although erection is normally the responsibility of the steel contractor, the design engineer can help in complex cases by scheduling the bracing that must be supplied as the construction is in progress. Alternatively, the contractor may be required to submit erection procedure plans to the engineer for approval.

Even after the main frames and members are successfully placed in the structure, failures have occasionally occurred because of the haste with which construction of main framing has proceeded without attention to the cross bracing that may be planned for the final structure in the planes of the walls and roof. After permanent bracing, roof, and walls are in place, the wind-load resistance of the structure will be greatly increased.

In summary, it may be said that construction failures are usually caused by lack of three-dimensional or "space frame" stability and that many more failures occur during erection than during service of the finished structures.

1.12 SERVICE AND MAINTENANCE REQUIREMENTS

The engineer, together with the architect and special consultants on such matters as heating, lighting, and ventilation, should give careful attention to the way in which the utility of the structure may be affected by the engineering design. Especially in an industrial building, structural design must be conditioned by the service functions of the structure.

Inadequate initial planning regarding the service which the structure is to perform will inevitably result in revisions in layout and corresponding costly design and material order changes before the structure is complete. It is obvious also that the locations of the electric wiring, heating ducts, and other service ducts for water, gas, chemicals, and so forth, as well as the locations of all special items of equipment, must be carefully predetermined, as all affect and are affected by the structural design.

Another service requirement of concern to the engineer is the desired life of the structure, together with consideration as to any special problems of corrosion that may exist due to atmospheric conditions, humidity, and so on. By proper design the engineer should avoid pockets where dirt and water may collect and should provide access to all parts of the structure that will require repeated painting and inspection during its life. Under adverse conditions, when maintenance cannot be assured, an extra thickness of metal may be furnished to allow for corrosion. Special corrosion-resistant steels are available, and another alternative is the use of *weathering steels,* which require no paint and develop a surface oxide that resists corrosion and presents a pleasing burnt-brown color.

During the useful life of an industrial plant, changes in processing procedures or even a complete change in use may come about. Thus, along with all necessary attention to special service requirements, an effort should be made to incorporate flexibility with respect to possible future alterations. The use of temporary interior partition walls is an example of such flexibility with respect to future change.

A structure should be designed to provide a life consistent with the buyer's wishes. A structure that is to last 100 years will be of quite different construction than one designed to survive for only a few years. The choice of materials used in construction may be affected; but even if both structures were of steel, there would need to be different consideration given to the problems of permissible stress, load evaluation, corrosion, painting, and other matters of upkeep for the two different life expectancies. The use of closed tube or box sections may materially reduce painting maintenance costs and be more justified in a long-lived structure than in a temporary one. Consideration under this category should also be given to fireproofing and fire protection. The difference in cost of fire insurance over the life of the structure must be weighed against the difference in initial cost between various degrees of fireproofing, assuming, however, that safety to human life is not an overriding consideration.

REFERENCES

1.1. *Specifications for the Design, Fabrication and Erection of Structural Steel for Buildings,*† American Institute of Steel Construction (1978).

1.2. *Manual of Steel Construction,*† 8th ed. American Institute of Steel Construction (1980) (includes Ref. 1.1).

1.3. *Structural Steel Detailing.* American Institute of Steel Construction (1971).

1.4. *Specification for the Design of Cold-Formed Steel Structural Members.* American Iron and Steel Institute (1968).

1.5. WINTER, GEORGE. *Commentary on 1968 Edition of the Specification for the Design of Cold-Formed Steel Structural Members.* American Iron and Steel Institute (1970).

†A mandatory supplement for the complete use of this book. Reference 1.1 will be referred to herein simply as AISCS; Ref. 1.2 will be referred to as AISCM. Although the 8th edition of the AISCM will be needed for problem work, the 7th edition, if owned, should be retained for additional reference material that it may contain.

1.6. JOHNSTON, B. G., ed., *The Structural Stability Research Council Guide to Stability Design Criteria for Metal Structures*, 3rd ed. John Wiley & Sons, Inc., New York (1976).

1.7. BROCKENBROUGH, R. L., and JOHNSTON, B. G., *The USS Steel Design Manual*. The United States Steel Corporation, rev. ed. (1974).

1.8. *Cold-Formed Steel Design Manual*. American Iron and Steel Institute (1977).

2

TENSION MEMBERS

2.1 INTRODUCTION

The most efficient way to use structural steel is in a tension member, that is, one that transmits "pull" between two points in a structure. Of course, if under certain load conditions the stress reverses in the member and becomes compression, the member must be designed both as a tension member and a column, and the efficiency is lost.

To make all the material in the tension member fully effective, the end connections must be designed to be stronger than the body of the member. If overloaded to failure, such a tension member will not only reach the yield stress but go above this level up to the ultimate strength of the material. In so doing it can absorb a great deal more energy per unit weight of material than any other type of member. This is an important consideration if impact or dynamic loads are a possibility. Beams and columns do not utilize material at full efficiency for two reasons: (1) metal failure is localized at highly stressed locations, and (2) some type of buckling failure always occurs at or below the yield stress, and the ultimate tensile strength of the material can never be reached.

Four types of tension member that can achieve high efficiency are illustrated in Fig. 2.1, showing (a) the wire rope or cable with socketed ends in which the use of cold-drawn steel wires having tensile strengths up to 150 ksi (or more, in special applications) provide the greatest strength/weight ratio available in the use of steel; (b) the simple round rod with threaded upset ends; (c) the eyebar, with forged ends for pin connections that are stronger

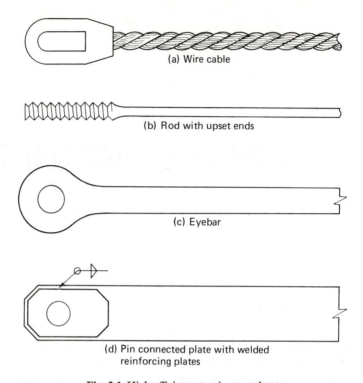

(a) Wire cable

(b) Rod with upset ends

(c) Eyebar

(d) Pin connected plate with welded
    reinforcing plates

Fig. 2.1 High-efficiency tension members.

than the body of the bar; and (d) the pin-connected plate with welded re-
inforcing plates at the ends.

In contrast to the foregoing, a tension member that can fail within its end
connection before yielding in the body of the member will absorb little energy
before failure—possibly less than 1% of the capacity it would have with
uniform yielding throughout its length. Regardless of where failure under
static load might occur, the tension member and its end connections should
be designed to guard against fatigue failure if alternate loading and unloading
is to be expected for a large number of repetitions.

Because of their efficiency, and because buckling is not a problem, tension
members make more advantageous use of the higher-strength steels than any
other type member.

No structural member is perfectly straight, and an intended axial force
will never act precisely along the longitudinal axis. As a result, there are
always "accidental" bending moments in such a member. In a column, as
illustrated in Fig. 2.2(a), these bending moments cause added deflection,
which further increases or "amplifies" both the deflection and the bending
moment caused thereby, equal to the product of the axial load and the
deflection.

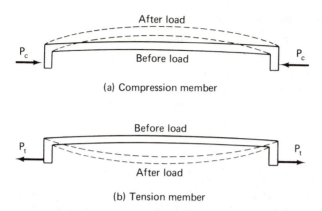

Fig. 2.2 Deflection of eccentrically loaded compression and tension members.

In an initially curved and eccentrically loaded tension member [Fig. 2.2(b)] the member tends to straighten, and the bending moments are reduced everywhere except at the end. Thus for very small accidental curvatures and end eccentricities, the additional tension stress induced by bending can usually be neglected unless design for repeated load is required.

2.2 TYPES OF TENSION MEMBERS

Four efficient tension-member types have been illustrated in Fig. 2.1. In addition, structural shapes and builtup members may be used, especially in trusses where tension and compression members must frame into a common joint, as shown in Fig. 2.3.

Fig. 2.3 Tension members (T) and compression member (C) entering lower chord joint of a truss.

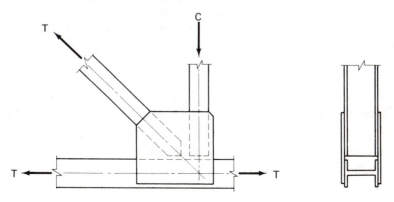

a. Wire Ropes and Cables

A cable is defined as a flexible tension member consisting of one or more groups of wires, strands, or ropes. A strand is formed by wires laid helically about a center wire so as to produce a symmetrical section; a wire rope is a plurality of strands laid helically around a central core that is composed of a strand, fabric core, or another wire strand, as illustrated in Fig. 2.4.

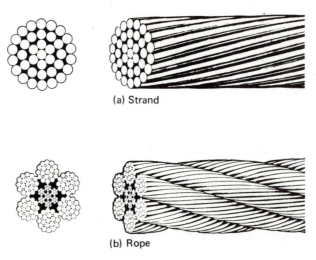

(a) Strand

(b) Rope

Fig. 2.4 Wire strand and wire rope, from U.S. Steel Corporation wire rope catalogue, by permission.

Wire cables are finding increasing use in structural steel design, and have been used both as primary and secondary supporting members in a wide variety of structures, including suspension bridges, prestressed concrete members, guyed towers, and wide-span roof structures. In roof construction the cables may radiate outward from a central tower, or may run radially inward from an outer compression ring, as illustrated in Fig. 2.5. The major U.S. steel producers issue catalogues that provide extensive design information and illustrations of the use of cables in roof structures.

b. Rods and Bars

The simplest tension member is the square or round rod. Round bars with threaded ends are less costly than bars with upset ends† [Fig. 2.1(b)], but have

---

†Upset ends were originally formed by forging. Currently, the threaded end segments may be made from a larger-diameter rod than the center segment and the three parts are then butt-welded together.

Fig. 2.5 Wire strand roof support system. (Courtesy American Institute of Steel Construction.)

certain disadvantages. Failure under impact overload or repeated load is apt to occur in the threaded portion. Bars with upset ends yield throughout their entire length and are recommended for the design of diagonal bracing for simple tower structures in earthquake regions. Large-diameter bars with threaded ends should be used with caution, because the lateral contraction in the diameter of the bar as yielding commences in the threaded portion may result in sufficient loss of thread-bearing area to result in failure by stripping of the threads before fully developing the maximum desired strength.

To guard against rods loosening after overloading, provision should be made for tightening at the ends of the member or by means of a turnbuckle between the ends of a two-piece member.

Round bars are frequently grouted into holes in rock formations to stabilize tunnel liners or retaining walls. They are also useful in reducing and restraining movement, such as a cracked machinery pedestal or spreading walls in old masonry structures.

In designing a rod with upset ends, the average stress on the area at the root of the thread should be less than the stress in the body of the bar. This will ensure yielding in the body of the bar if there is severe overload, as in an earthquake. If there is misalignment or bending in a tie rod, the use of upset ends offers the additional advantage that any added stress due to bending is greatest in the main body of the bar, which is most flexible and able to adjust to such a condition.

Areas of rods and bars are tabulated in the AISCM as are dimensions of threads, turnbuckles, clevises, and sleeve nuts for a wide variety of rod diameters.

c. Eyebars and Pin-Connected Plates

Eyebars and pin-connected plates [Fig. 2.1(c) and (d)] are used in a variety of special situations. Examples include the transfer of tensile load from a wire rope or cable to a structural steel assemblage or to an anchorage, as in the case of a suspension bridge. The use of eyebars as tension members in a modern long-span bridge is illustrated in Fig. 2.6.

If failure were to occur in the head of the eyebar at the connection, tests have demonstrated that it would be one of the following types:

1. Fracture behind the pin in a direction parallel to the axis of the bar. This type of failure will occur if insufficient edge distance behind the pin is provided [Fig. 2.7(a)].

2. Failure in the net section through the pin transverse to the axis of the bar. This type of failure will occur if the gross area of the main section of the bar is equal to or greater than the net section through the pin hole [Fig. 2.7(b)].

3. Failure by *dishing*. This is an inelastic lateral stability type of failure, which will occur if the width/thickness ratio behind the pin is too great. Dishing failure is akin to the lateral instability of a short deep beam [Fig. 2.7(c)].

Since it is desirable to ensure that general yielding and ultimate failure occur in the main body of the bar rather than at the end, all specifications provide dimensional requirements that prevent failure of the types enumerated above. When several eyebars come together or connect at the same pin with packing and external nuts to prevent spreading of the package, there will be no need to restrict the width/thickness ratio of the eyebar, since dishing will be prevented by the lateral constraints provided. The required proportion for eyebars is similar in the various specifications.

In general, connection details are treated in Chapter 6; but in the case of eyebars or pin-connected plates, the design of end-connecting details, except

Fig. 2.6 Eyebar tension members in the Brent Spence Bridge over the Ohio River.

for the proportioning of the pin itself, is related to the design of the whole member and will be covered in this chapter. The AISCS, Sec. 1.14.5, specifies proportions that provide for a balanced design with respect to the various possible modes of failure. It should be studied in detail by reference to Example 2.1.

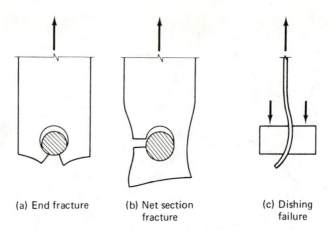

(a) End fracture   (b) Net section   (c) Dishing
                          fracture   failure

Fig. 2.7 Various failure modes of a pin-connected plate.

d. Structural Shapes and Built-up Members

Structural shapes and built-up members are used when rigidity is required in a tension member, to resist small lateral loads, or when reversal of load may subject the member to alternate compression and tension, as in a truss diagonal near the center of a span. The most commonly used shapes are the angle, tee, and W, S, or M shapes, as shown in Fig. 2.8. For exposed use, to minimize wind load, the pipe section may be favored, although the end connections present a problem in truss construction. Built-up members are

Fig. 2.8 Structural shapes as tension members.

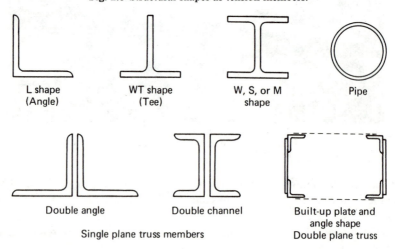

L shape   WT shape   W, S, or M   Pipe
(Angle)   (Tee)   shape

Double angle   Double channel   Built-up plate and
                                               angle shape
Single plane truss members   Double plane truss

formed by connecting two or more structural shapes with separators, battens, lacing, or continuous plates, so that they will work together as a unit, as shown in Fig. 2.8. The angle and channel members, as shown, may be used in single-plane truss construction connected to end gusset plates with rivets, bolts, or welds. The built-up open-box shape, as shown (the dashed lines indicate lacing or battens), is suitable for double-plane truss construction and is conveniently bolted, riveted, or welded between two gusset plates at each end connection. W, M, or S shapes are especially suited to double-plane welded truss construction.

Although end connections are of paramount importance in the design of a complete tension member, this topic, except for the eyebar, will be covered in Chapter 6.

A nonmandatory clause, Sec. 1.8.4. of the AISCS, recommends a maximum slenderness ratio (l/r) of 240 for main members and 300 for lateral bracing and other secondary members. Rods and cables are excepted from these limitations.

2.3 ALLOWABLE TENSILE STRESS AND EFFECTIVE NET AREA

Axially loaded tension members are proportioned so that the nominal, or average, stress shall not exceed the specified allowable tensile stress as defined by the specification as a proportionate part of either the yield stress or ultimate tensile strength, whichever is critical. The nominal tensile stress f_t is simply the anticipated axial design load P divided by the area A of the member at the particular location under consideration. In riveted and bolted connections A is taken as the gross area in the check for safety against yield and is taken as the *effective* net area in the design check for safety with respect to ultimate strength at failure under static or essentially nonrepetitive loads. In the case of pin-connected members, the actual net area at the pinhole is used to calculate f_t.

Although an effort should be made to reduce local concentration of stress due to changes in section, particularly at connecting welds, by use of smoothly tapered and gradual transitions, the local stresses are not usually added to the "nominal" stress. Tests to failure have shown that regions of local yield in a well-designed and properly fabricated tension member do not prevent the entire cross section from reaching the yield point and beyond, thus developing the full strength of the member before failure.

For buildings, the allowable tensile stress is specified by the AISCS, Sec. 1.5.1.1, as follows:

1. Allowable tensile stress F_t on the gross area or effective net area of the tension member, except for pin holes:

On the gross area:

$$F_t = 0.6F_y$$

On the effective net area:

$$F_t = 0.5F_u$$

where

F_y = specified minimum yield stress, ksi

F_u = specified minimum tensile strength, ksi

The effective net area $A_e = C_t A_n$, where A_n is the actual net area and C_t is a reduction factor dependent on the type of shape and arrangement of the rivet or bolt pattern, as stipulated by AISCS, Sec. 1.14.2.2. The actual net area A_n, in the case of a chain of holes extending across an element, is determined by an empirical rule as defined in AISCS, Sec. 1.14.2.1: "the net width of the part shall be obtained by deducting from the gross width the sum of the diameters of all the holes in the chain, and adding, for each gage space in the chain, the quantity

$$\frac{s^2}{4g}$$

where

s = longitudinal spacing (pitch) of any two consecutive holes

g = transverse spacing (gage) of the same two holes

The critical net area A_n of the part is obtained from that chain which gives the least net width."

The following example illustrates the application of the foregoing rule for net width calculation. In the example, a plate, $\frac{3}{4}$ by 10, is in tension and is attached to another plate by means of 14 high-strength $\frac{3}{4}\ \phi$ bolts. The connection strength of the bolts will not be evaluated, that being a topic considered in Chapter 6. Example 2.1 considers only the strength as determined by net section, either through line *abde*, deducting two holes, or through *abcde*, deducting three holes and adding, by the rule above, the value of $s^2/4g$ as determined by *bc* and *cd*.

Example 2.1

Determine the allowable tensile force as determined by the following hole pattern:

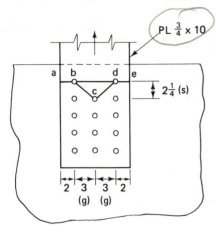

Steel: ASTM A36

$F_y = 36$ ksi

$F_u = 58$ ksi

$F_t = 0.6 \times 36 = 22^*$ ksi (gross area)

$F_t = 0.5 \times 58 = 29$ ksi (net area)

Holes (standard) for $\frac{3}{4}$ bolts, will have a $\frac{13}{16}$ diam. (AISC Sec. 1.23.4.1)

Deduction for hole diam. = hole diam. plus $\frac{1}{16} = \frac{7}{8}$ (AISC Sec. 1.14.4)

Gross area = $\left(\frac{3}{4}\right) \times 10 = 7.50$ in.^{2}

---

* 22 is rounded off from 21.6 as allowed by the AISCS.

Solution

The effective net area of this plate is equal the actual net area (AISCS, Sec. 1.14.2.3). The net area is the lesser of

$$0.75 \times (10 - 2 \times \tfrac{7}{8}) = 6.19 \text{ in.}^2 \text{ (line } abde)$$

$$0.75 \times \left(10 - 3 \times \tfrac{7}{8} + \frac{2 \times 2.25^2}{4 \times 3}\right) = 6.16 \text{ in.}^2 \text{ (line } abcde)$$

The allowable tensile force for A36 steel is the lesser of

$$22 \times 7.5 = 165 \text{ kips} \qquad \text{(gross section check)}$$

$$29 \times 6.16 = 178.6 \text{ kips} \qquad \text{(net section check)}$$

The effective net area is not to be taken as more than 85% of the gross area (AISCS, Sec. 1.14.2.3):

$$0.85 \times 7.5 = 6.38 \text{ in.}^2 > 6.16 \qquad \text{OK}$$

Assuming a capacity of 165 kips, the possibility exists that the full net section with three holes deducted might control the strength after the first two fasteners have taken their share of the load, the remaining load being equal to

$$\tfrac{12}{14} \times 165 = 141.4 \text{ kips}$$

The net area, three holes deducted, is

$$0.75 \times (10 - 3 \times \tfrac{7}{8}) = 5.53 \text{ in.}^2$$

The capacity at the line through c is

$$29 \times 5.53 = 160.4 \text{ kips} > 141.4 \qquad \text{OK}$$

When a structural shape whose elements are not in one plane is loaded in tension and attached by some, but not all, of its segments by rivets or bolts, the effective net area is less than the actual net area. The procedure for determining net width is the same as that illustrated in Example 2.1. Reference should be made to AISCS, Sec. 1.14.2.2, for the amount of reduction of net area for various sections and different fastener arrangements. The topic of shear lag, referred to in the AISCS Commentary Sec. 1.14.2.2, is beyond the scope of this text.

Table 2.1†

Allowable Stress as a Function of F_y

F_y (ksi)	Allowable Stress (ksi)					
	$0.4F_y$	$0.45F_y$	$0.60F_y$	$0.66F_y$	$0.75F_y$	$0.90F_y$
36	14.5	16.2	22.0	24.0	27.0	32.4
40	16.0	18.0	24.0	26.4	30.0	36.0
42	16.8	18.9	25.2	27.7	31.5	37.8
46	18.4	20.7	27.6	30.4	34.5	41.4
50	20.0	22.5	30.0	33.0	37.5	45.0
60	24.0	27.0	36.0	39.6	45.0	54.0
65	26.0	29.3	39.0	42.9	48.8	58.5
90	36.0	40.5	54.0	*	*	81.0
100	40.0	45.0	60.0	*	*	90.0

*These steels not used in plastic design. See Chapter 8.

Table 2.2†

Allowable Stress as a Function of F_u

F_u (ksi)	Allowable Stress (ksi)				
	$0.5F_u$	$1.5F_u$	$0.33F_u$	$0.17F_u$	$0.22F_u$
58	29.0	87.0	19.1	9.9	12.8
60	30.0	90.0	19.8	10.2	13.2
63	31.5	94.5	20.8	10.7	13.9
65	32.5	97.5	21.5	11.1	14.3
67	33.5	101.0	22.1	11.4	14.7
70	35.0	105.0	23.1	11.9	15.4
75	37.5	113.0	24.8	12.8	16.5
80	40.0	120.0	26.4	13.6	17.6
100	50.0	150.0	33.0	17.0	22.0
110	55.0	165.0	36.3	18.7	24.2

†Adapted from Tables 1 and 2, AISCS, Appendix A.

2. The allowable tensile stress F_t on the net section at pin holes in eyebars, pin-connected plates, or built-up members is

$$F_t = 0.45F_y$$

In summary, the allowable tensile stresses, F_t, are tabulated for the various yield stresses and tensile strengths in Tables 2.1 and 2.2, respectively, as adapted from more extensive tables in the AISCS. Numerical values are rounded off as accepted by AISC and the tables include listings both for Chapter 2 and later chapters in the text.

2.4 DESIGN FOR REPEATED LOAD

When load is repeatedly applied and removed, with the number of repetitions running into the many thousands or up into the millions, metal may develop cracks that eventually may spread to the point where they cause *fatigue failure* of the member. Fatigue cracks are most apt to occur when the repeated load is primarily tension. Local stress concentrations increase the susceptibility to fatigue failure. Such concentrations may be due to poorly made welds, small holes, and rough or damaged edges resulting from the fabrication processes of shearing, punching, or poor-quality oxygen cutting. The fatigue strength of the higher-strength steels has not been shown to be appreciably greater than the commonly used A36 grade structural steel with a yield stress of 36 ksi.

In 1969 the AISCS introduced a simplified approach to design for repeated load, as presented in Appendix B of the AISCS. The unique feature of the AISCS approach to repeated load design is its use of the expected *stress range* as the controlling design criterion. Stress range is the algebraic difference between maximum and minimum stress to be expected in any one cycle of loading. Thus the following two cases each have the same stress range of 16 ksi:

Maximum stress	Minimum stress
20 ksi tension	4 ksi tension
4 ksi tension	−12 ksi compression

The permissible stress range is a function of (1) loading condition, and (2) stress category.

Loading conditions are defined and tabulated in Table B1, page 84, of the AISCS, and are determined according to the anticipated number of loading cycles to be used as a basis for design. If less than 20,000 cycles, no consideration need be given to repeated load, but at successive lower limits of 20,000, 100,000, 500,000, and 2,000,000 cycles, respectively, load conditions 1, 2, 3, and 4 are established.

The stress categories, ranging from A to F with increasing severity of the local stress raiser, are tabulated and defined in Table B2 of the AISCS (pages 85 to 87). After establishing the loading condition and the stress category, the allowable stress range is read from AISCS, Table B3, page 87.

Members in conventional buildings usually do not need to be designed for repeated load because the number of repetitions of maximum load is usually less than 20,000. However, crane runway girders and supporting structures for machinery and equipment do require the consideration of fatigue.

2.5 FLOWCHART

Flowchart 2.1. *Tension-Member Selection*

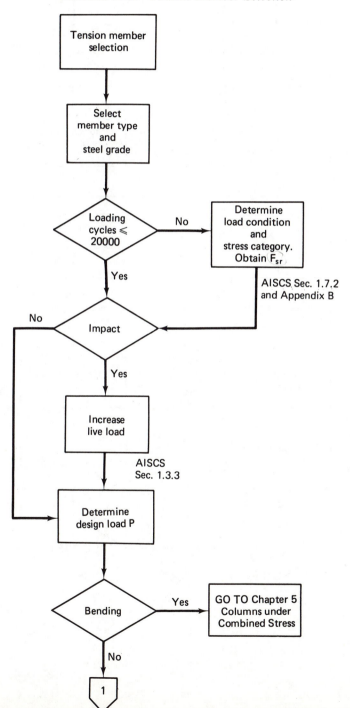

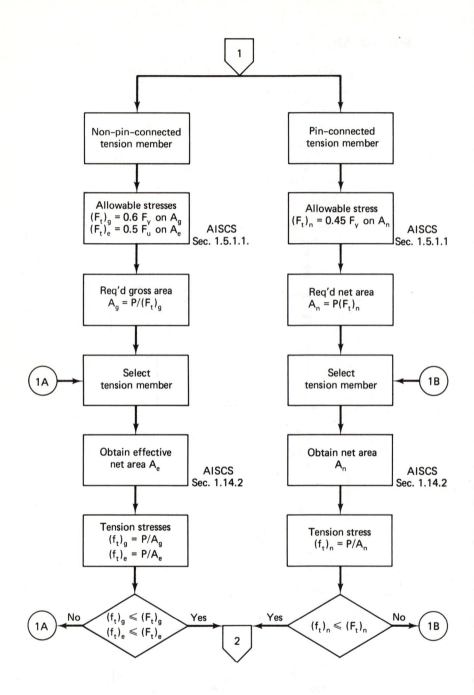

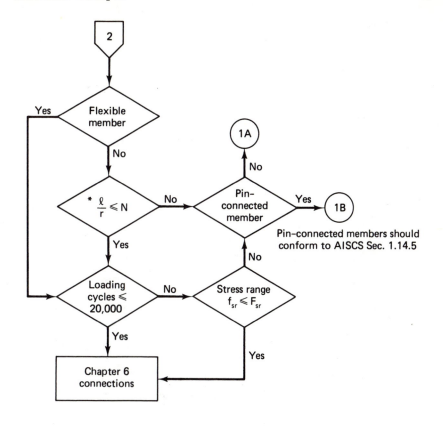

* AISCS Sec. 1.8.4:
N = 240 for main members.
N = 300 for lateral bracing members
and other secondary members.

2.6 ILLUSTRATIVE EXAMPLES

Example 2.2

A floorbeam suspender is stressed in tension by a dead load of 30 kips and a live load of 40 kips. The full live load will be repeated less than 20,000 times. Select a round steel rod, using upset ends, to satisfy the AISCS. Use quenched and tempered alloy steel with $F_y = 100$ ksi.

Solution

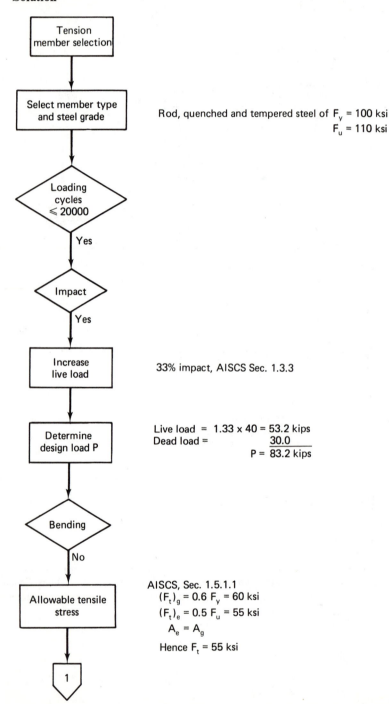

Select member type and steel grade → Rod, quenched and tempered steel of $F_y = 100$ ksi
$F_u = 110$ ksi

Increase live load → 33% impact, AISCS Sec. 1.3.3

Determine design load P →
Live load $=$ $1.33 \times 40 = 53.2$ kips
Dead load $=$ $\underline{\hspace{1cm}30.0}$
$P = \overline{83.2}$ kips

Allowable tensile stress →
AISCS, Sec. 1.5.1.1
$(F_t)_g = 0.6 F_y = 60$ ksi
$(F_t)_e = 0.5 F_u = 55$ ksi
$A_e = A_g$
Hence $F_t = 55$ ksi

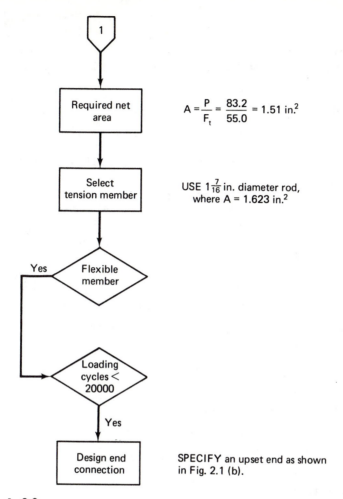

Required net area

$$A = \frac{P}{F_t} = \frac{83.2}{55.0} = 1.51 \ \text{in.}^2$$

Select tension member

USE $1\frac{7}{16}$ in. diameter rod, where $A = 1.623$ in.^{2}

Flexible member — Yes

Loading cycles $<$ 20000 — Yes

Design end connection

SPECIFY an upset end as shown in Fig. 2.1 (b).

Example 2.3

Design an eyebar to carry a tensile load of 600 kips (less than 20,000 repetition of load). Use steel with $F_y = 50$ ksi.

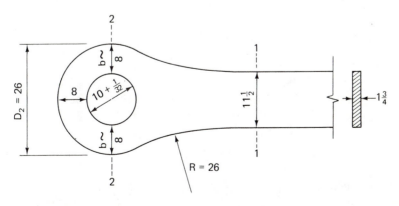

Select ASTM A572 steel.

Given $P_{(\text{ten.})} = 600$ kips, $F_y = 50$ ksi, $F_u = 65$ ksi, $(F_{t_1})_g = 0.6F_y = 30$ ksi, $(F_{t_1})_e = 0.5F_u = 32.5$ ksi, and $A_e = A_g$.

Solution

(1.5.1.1)† $F_{t_1} = 0.6(50) = 30$ ksi $F_{t_2} = 0.45(50) = 22.5$ ksi

$$A_1 = \frac{600}{30} = 20 \text{ in.}^2 \quad \text{Use PL } 1\tfrac{3}{4} \times 11\tfrac{1}{2} \; (A_1 = 20.12 \text{ in.}^2)$$

(1.14.5)† $\left(\dfrac{b}{t}\right)_1 = \dfrac{11.5}{1.75} = 6.57 < 8$ OK

$$A_{2(\text{min.})} = \frac{600}{22.5} = 26.6 \text{ in.}^2$$

or $1.33 \times A_1 = 26.8$ in.^{2} ⟵ governs

$$b \text{ at } 2\text{–}2 = \frac{26.8}{2 \times 1.75} = 7.67 \text{ in.} \quad \text{Use 8 in. (approx.)}$$

$A_{2(\text{act.})} = 2 \times 8 \times 1.75 = 28.0$ in.^{2} $< 1.5A_1 = 30.2$ in.^{2} OK

diam. pin $\geq \tfrac{7}{8}(11.5) = 10.08$ in.   Use 10-in. pin

diam. hole $= 10 + \tfrac{1}{32} = 10\tfrac{1}{32}$

$D_2 = 10\tfrac{1}{32} + 2 \times 8 = 26\tfrac{1}{32}$ Use 26 in. $(b \approx 8 \text{ in.})$

$R > D_2 = 26\tfrac{1}{32}$ Use 28 in.

(1.5.1.5.1)†   Check the bearing:

$$F_p = 0.9F_y = 0.9 \times 50 = 45 \text{ ksi}$$

$$f_p = \frac{600}{10 \times 1.75} = 34.3 \text{ ksi} < 45 \quad \text{OK}$$

Example 2.4

A tension member of a roof truss has a length of 25 ft and is stressed in tension by a dead load of 40 kips and a live load of 60 kips. The tension member is a main

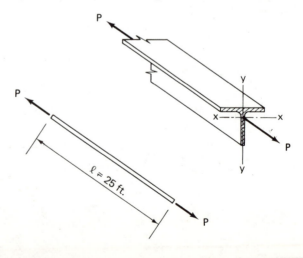

†AISCS sections.

member and needs some amount of rigidity. Select a single structural tee to satisfy the AISCS. Use A36 steel.

Solution

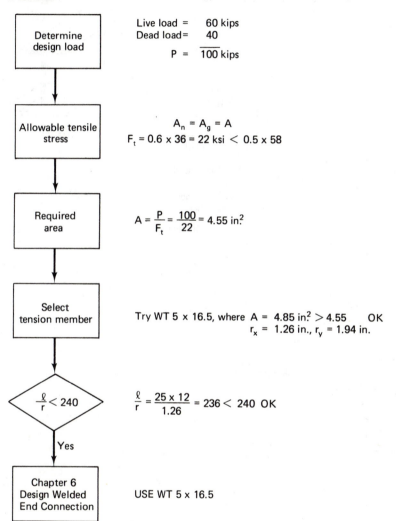

Determine design load

Live load = 60 kips
Dead load = 40
P = $\overline{100}$ kips

Allowable tensile stress

$A_n = A_g = A$
$F_t = 0.6 \times 36 = 22$ ksi $< 0.5 \times 58$

Required area

$A = \dfrac{P}{F_t} = \dfrac{100}{22} = 4.55$ in.^{2}

Select tension member

Try WT 5 × 16.5, where A = 4.85 in.^{2} > 4.55   OK
$r_x = 1.26$ in., $r_y = 1.94$ in.

$\dfrac{\ell}{r} < 240$

$\dfrac{\ell}{r} = \dfrac{25 \times 12}{1.26} = 236 < 240$ OK

Yes

Chapter 6 Design Welded End Connection

USE WT 5 × 16.5

Example 2.5

Same as Example 2.4, but an additional axial tension of 45 kips produced by wind should be considered if it governs the design.

Solution

Design load

Dead load = 40 kips
Live load   = 60
                 $\overline{100}$
Wind load = 45 kips

Percentage of wind load to the sum of the design dead load and live load:

$$\frac{45}{100} \times 100 = 45\% > 33.3 \quad \text{AISCS Sec. 1.5.6}$$

Therefore, wind load governs the design of the tension member. Then,

$$P = 100 + 45 = 145 \text{ kips}$$

Allowable tensile stress

$$F_t = 1.333 \times 0.6F_y = 1.333 \times 22 \quad \text{(For wind load included)}$$

Area required

$$A = \frac{145}{1.333 \times 22} = 4.94 \text{ in.}^2$$

Select tension member

Try W 6 x 20, where A = 5.87 in.^{2} > 4.94   OK
r_x = 2.66 in., r_y = 1.50 in.

$$\frac{\ell}{r} < 240$$

$$\frac{\ell}{r} = \frac{25 \times 12}{1.50} = 200 < 240 \quad \text{OK}$$

Yes

Chapter 6 Design Welded End Connection

USE W 6 x 20

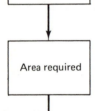
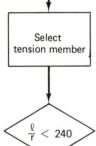

Example 2.6

Same as Example 2.4, except that the live load of 60 kips may be repeated 300,000 times and during each cycle the member is subjected to 10-kips compression. Welded end connection with fillet welds is similar to Case 17 in Appendix B of AISCS.

Solution

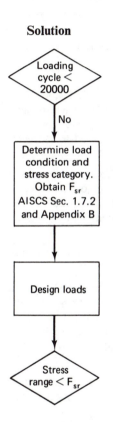

Loading condition 2 for 300,000 loading cycles. Illustrative Example no. 17 for fillet welded connection.
Stress category E

Therefore, allowable stress range
$$F_{sr} = 12.5 \text{ ksi}$$

Tension:
  Dead load + 40
  Live load   + 60 P_t = +100 kips (max.)

Compression:
  Dead load + 40
  Live load   − 10 P_t = +30 kips (min.)

WT 5 x 16.5 is the selected tension member as Example 2.3 for the design dead and live loads of 100 kips.

Then check stress range for this fatigue loading condition:

Actual stresses:

$$\text{Max.} \quad f_t = \frac{P_t}{A} = \frac{100}{4.85} = 20.6 \text{ ksi}$$

$$\text{Min.} \quad f_t = \frac{P_t}{A} = \frac{30}{4.85} = 6.18 \text{ ksi}$$

$$\text{Actual stress range} = 20.6 - 6.18$$
$$= 14.42 \text{ ksi} > F_{sr} \quad \text{NG}$$

Try WT 5 x 19.5 (Next larger size)   A = 5.73 in.²
New stress range, by proportion of areas

$$= \frac{4.85}{5.73} \times 14.42 = 12.20 < F_{sr} \quad \text{OK}$$

PROBLEMS†

2.1. In place of the upset round rod chosen in Example 2.2, select a rectangular bar for the same design requirements. Assume welded end connections so as to make the full cross section of the bar available as net section in tension.

2.2. Same as Problem 2.1, but with high-strength bolted end connections, using a single line of $\frac{3}{4}$-in.-diameter bolts. Refer to AISCS, Secs. 1.14.2 and 1.14.4, for guidance in determining the net section of the bar.

2.3. Redesign an eyebar to meet the requirements of Example 2.3, but with the steel yield point changed from 50 to 42 ksi, using ASTM A242 steel with $F_u = 63$ ksi.

2.4. Design a tension member to meet the requirements of Example 2.4, but with the structural tee section replaced by two angles, with long legs back to back. Long legs are to be separated $\frac{3}{8}$ in. for end connections to gusset plates. In determining effective net section, assume a single line of holes for $\frac{7}{8}$-in.-diameter bolts through long sides of angles and refer to AISCS, Sec. 1.14.2.

2.5. Same as Problem 2.4, except that the live load of 60 kips may be repeated 600,000 times, and during each cycle the member may be subjected to a compression of 35 kips. The high-strength bolted end connection may be assumed to be equivalent, in fatigue resistance, to Case 8 in Appendix B of AISCS. Example 2.6, should be studied as a guide.

2.6. Same as Problem 2.4, but with the addition of a 40-kip tension force due to wind. Refer to Example 2.5.

2.7. The $1\frac{7}{16} \phi$ bar was chosen in Example 2.2 without consideration of repeated load; that is, the repetitions of load were assumed to be less than 20,000. Ruling out the possibility of failure in the end threads, show that the bar is capable of withstanding 300,000 repetitions of live load (without impact). For repeated load, the round bar may be assumed as equivalent to Case 2 in Appendix B of AISCS.

2.8. Referring again to Example 2.2, assume that the thread root area in the upset end of the rod is 1.90 in.². Assume that this puts the end in repeated load Category F of Table B3 of AISCS. Is the threaded end adequate for 300,000 repetitions of live load, alone, excluding impact? If not, what change would you recommend? Such a change could involve a different material, different type of member, or a different end connection.

†All problems in this and succeeding chapters are to be done in accord with AISCS. Note that the term "example" refers always to the illustrative examples in the body of the chapter and the term "problem" refers to problems at the end of the chapters. Shapes and bars chosen in problem solutions should be "available" as designated by the AISCM.

3

BEAMS

3.1 INTRODUCTION

Beams support loads that are applied at right angles (transverse) to the longitudinal axis of the member. Such loads are usually downward as illustrated in Fig. 3.1(a). The beam carries the loads to its supports, which may consist of the bearing walls, columns, or other beams into which it frames. At the supports the upward "reactions" have a total magnitude equal to the weight of the beam plus the applied loads P. Since the weight of the beam is not known until after it is designed, the design starts with a preliminary weight estimate that is subject to later revision.

Imagine a free-body diagram of the left portion of the beam [Fig. 3.1(b)] with bending moment (M) and the shear (V) necessary at the cut section to provide static equilibrium. The problem of beam design consists mainly in providing enough bending strength and enough shear strength at every location in the span. For short spans, it is most economical to use a single-beam cross section throughout the span, and in such case only the maximum values of bending moment and shear need to be determined.

A *simple beam* [Fig. 3.1(a)] is supported vertically at each end with little or no rotational restraint, and downward loads cause positive bending moment throughout the span. The top part of the beam shortens, due to compression, and the bottom part of the beam lengthens, due to tension [Fig. 3.1(d)]. The most common rolled steel beam cross section, shown in Fig. 3.1(c), is called the W shape, with much of the material in the top and bottom flange, where it is most effective in resisting bending moment. The

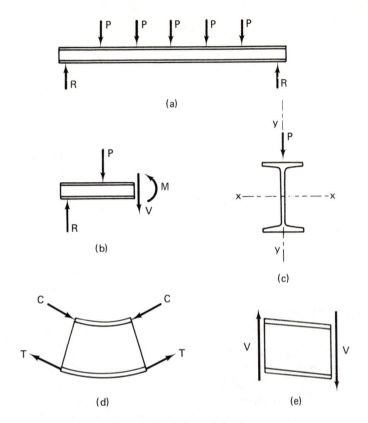

Fig. 3.1 Simple beam behavior.

web of the beam supplies most of the shear resistance and in so doing is slightly distorted, as shown in Fig. 3.1(e). The contribution of this distortion to beam deflection is usually neglected. The bending moment causes curvature of the beam axis, concave upward, as shown in Fig. 3.1(d) for positive moment, concave downward for negative moment. The deflection of beams is calculated on the assumption that it is entirely caused by the curvature due to bending moment.

Standard AISC nomenclature pertaining to the W (wide-flange) hot-rolled steel beams is illustrated in Fig. 3.2.

The reader should gain a familiarity with information in the AISCM relative to rolled shapes, reading the descriptive material and scanning the tabular material, which includes:

1. Discussion and tabular material relative to shape selection, designation, dimensions, availability, size groupings, principal producers, and proper manner of shape designation.

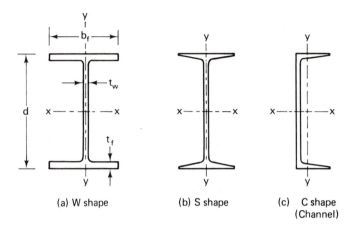

(a) W shape (b) S shape (c) C shape
(Channel)

Fig. 3.2 Nomenclature pertaining to a beam cross section, where x–x = principal major axis, strong axis of bending, weak plane; y–y = principal minor axis, weak axis of bending, strong plane.

2. Dimensions of shape cross sections for detailing.
3. Properties of shape cross sections for use in design calculations.
4. Standard mill practice in the rolling, cambering, and cutting of shapes, with corresponding dimensional tolerances.

A *plate girder* (see Chapter 7) is of such large depth and span that a rolled beam is not economically suitable—it is tailor made (built up out of plate material by use of welds, rivets, or bolts) to suit the particular span, clearance, and load requirements.

It is assumed that the reader is familiar with the analysis of shears and moments, with the drawing of corresponding shear and moment diagrams, and with the usual designation of support conditions. Various cases are illustrated in Fig. 3.3. At the top the loads and supports are shown for (a) a cantilever beam, (b) a simple beam with a cantilever overhang at the right end, and (c) a beam fixed at the left end and the same as (b) at the right end. In (c) the shears and moments between the fixed end and the simple support are *statically indeterminate;* that is, they cannot be determined by simple statics. Shear diagrams are shown on the second line, moment diagrams on the third. Although the calculation of shears and moments will be included in many of the illustrative examples, reference should be made to a text on strength of materials or elementary structural theory for additional information on these topics. For uniform or distributed loads the shear and moment diagrams are similar to those shown in Fig. 3.3, but the shear, since it changes with load, is a sloping line instead of a horizontal one, and the moment diagram is a continuous curve between reactions. The reader should review the

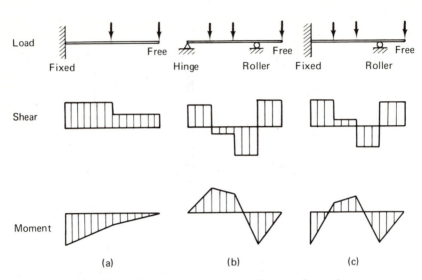

Fig. 3.3 Load, shear, and moment diagrams for various
beam situations.

mathematical relationships between load, shear, and bending moment as
found in texts on strength of materials or structural theory.

Beams usually are framed with other beams, or attached to a floor slab,
as shown in Fig. 3.4, so that the beam cannot move sideways and the beam is
forced to deflect vertically in the strong (y–y) plane (see Fig. 3.2).

Whenever a beam deflects in the plane in which it is loaded, the simple
theory of bending† may be used. The condition may be forced, as previously
mentioned, or it can occur naturally if the plane of the loads contains a
principal axis of the cross section. However, if the load is in the strong (y–y)
plane (see Fig. 3.2), the beam may need lateral support to prevent it from
buckling sideways; alternatively, the specifications provide for reduced
allowable loads if lateral support does not meet certain minimal requirements.
If loaded in the weak (x–x) plane (see Fig. 3.2), lateral buckling is no problem.
Sections that lack two axes of symmetry usually require more positive lateral
supports than does the W shape. For example, the laterally unsupported
channel member will twist if loaded through the centroidal axis, as shown in
Fig. 3.5(b), and requires restraint against both twist and lateral buckling.
The zee section does not twist but deflects at an angle to the plane of the loads
unless supported as shown in Fig. 3.5(c). An angle loaded as shown in Fig.
3.5(d) must be supported against both twist and lateral deflection. It is also
important to recognize that if the zee or angle section is used without lateral
support, the stress due to bending cannot be calculated by the simple beam

---

†The simple theory of bending will be reviewed briefly in Section 3.2.

Fig. 3.4 Beams framed to column and supporting permanent
metal forms for floor slab. (Courtesy Bethlehem Steel Corp.)

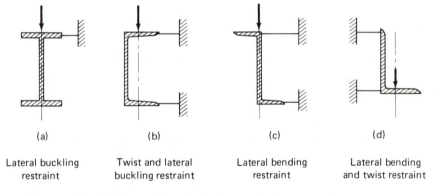

(a)	(b)	(c)	(d)
Lateral buckling	Twist and lateral	Lateral bending	Lateral bending
restraint	buckling restraint	restraint	and twist restraint

Fig. 3.5 Type of lateral restraint required to permit beam
selection by simple bending theory.

formula. Where lateral support is needed only to prevent lateral buckling
[Fig. 3.5(a)] there is no calculable stress in the lateral supports. In cases (b),
(c), and (d), however, there is a calculable stress in the lateral supporting
members, and thus a more clearly defined design problem exists.

Most beams are designed by simple bending theory. The design process
involves the calculation of the maximum bending moment and the selection

of a beam having an equal or greater bending moment resistance. The selection is then checked for maximum shear capacity, and the end connections or bearing support details are designed. A deflection check may also be required.

Some of the more complex beam design problems, such as general biaxial bending and combined bending and torsion, are treated in Chapter 10. An introduction to plastic design of continuous beams and frames is presented in Chapter 8. In plastic design the required design load is multiplied by a load factor to give the required ultimate collapse load, and the continuous beam or frame is chosen to have equal or greater ultimate load capacity. The *stress* due to bending is not calculated—at ultimate load in various sections of the beam it will be at or even slightly greater than the yield point. Fully continuous beams or frames are statically indeterminate in the elastic range, but the analysis problem becomes statically determinate when the ultimate strength is reached, another advantage for plastic design. However, allowable stress (elastic) design is customary and adequate for the design of statically determinate beams, such as were illustrated in Figs. 3.1 and 3.3(a) and (b). Elastic, or allowable-stress, design will be emphasized in this chapter, although a brief introduction to beam behavior in the inelastic range (Section 3.3) will be included, because it is essential to an understanding of specification modifications of allowable stresses as well as to the study of plastic design.

3.2   ELASTIC BENDING OF STEEL BEAMS

A knowledge of elementary beam theory as presented in texts on mechanics or strength of materials is an essential preliminary to the study of beam design. Figure 3.6 shows a *unit* length of beam imagined as cut out of the complete beam at any location along the beam. It is acted upon by bending moment M and shear V, positive as indicated, and is shown in its undeflected straight position before loading and in its deflected and bent position after loading. Note that Y and y, positive as shown, are used to signify two different distances: deflection of the beam axis and distance within the beam cross section from the centroid, respectively. The *curvature* of the beam, or change in slope per unit length of beam, is denoted by ϕ, and the unit longitudinal *strain*, or change in length per unit length, of a horizontal beam fiber is therefore equal to

$$\epsilon = \phi y \tag{3.1}$$

Since normal stress (f_b) is equal to the modulus of elasticity (E) multiplied by strain (ϵ), the stress due to bending is equal, by Eq. (3.1), to

$$f_b = E\phi y \tag{3.2}$$

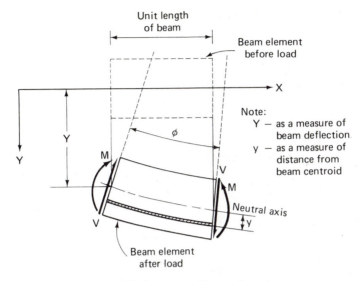

Fig. 3.6 Deformation of beam element.

Thus the stress due to bending of a beam is known if the curvature is known. This fact could be of interest to a wire manufacturer who wished to know the diameter of a drum or reel on which drawn wire could be wound without inducing any permanent bend. Suppose, for example, that a wire with a diameter of 0.10 in. is wound on a reel having a diameter of 60 in. The curvature of the wire is equal to a unit length (1 in.) divided by the reel radius (30 in.). Thus

$$\phi = \tfrac{1}{30}$$

The maximum strain in the wire, by Eq. (3.1), is

$$\epsilon = \tfrac{1}{30}(0.05) = 0.00167$$

The maximum stress, then, for steel wire, due to bending, for $E = 29,000$ ksi, is equal to

$$f_b = (29,000)(0.00167) = 48.3 \text{ ksi}$$

The stress of 48.3 ksi is greater than the yield point of carbon structural steel, but less than the elastic limit of most cold-drawn high-strength steel wires, for which the reel diameter would be satisfactory as it would not induce permanent bending deformation in the wire.

Equation (3.2) is convenient in the problem facing the wire manufacturer, but for the allowable stress design of steel beams the stress due to bending is usually calculated as a function of the bending moment, which is proportional to the curvature. The constant of proportionality between moment and curvature is EI, I being the moment of inertia of the cross section, as tabulated for all rolled sections in the AISCM. Thus the bending moment is

$$M = EI\phi \qquad (3.3)$$

Combining Eqs. (3.3) and (3.2), the formula for stress in terms of bending moment is obtained:

$$f_b = \frac{My}{I} \qquad (3.4)$$

Equation (3.4) is sometimes called the *beam equation*, and its use is restricted to the previously described simple bending theory. Initial beam selection is based on the maximum stress due to bending, for which $y = c$, where c is the maximum y distance from the centroidal (also neutral) axis of the beam to the extreme top or bottom fiber of the cross section. If the beam section is symmetrical about its x axis, c will be the same for the compression and tension extremities.

$$(f_b)_{max} = \frac{M_{max}c}{I} \qquad (3.5)$$

To expedite the design selection of a beam for maximum bending moment, I and c are combined into a single parameter, the *section modulus*, denoted by S and equal to I/c.

Equation (3.5) then becomes simply

$$(f_b)_{max} = \frac{M_{max}}{S} \qquad (3.6)$$

In allowable stress design the maximum stress due to bending (f_b) must be less than the allowable stress in bending (F_b) as specified by the AISCS; thus the required section modulus in the design of a beam is

$$S_{reqd} = \frac{M_{max}}{F_b} \qquad (3.7)$$

Perhaps the most commonly used table in the AISCM is the one that tabulates the section modulus values for all rolled shapes used as beams. The tabulation excludes shapes normally used as columns, which are primarily those with flange widths approximately equal to the depth of the section. The explanatory introduction to this tabulation in AISCM is recommended reading.

Unless the beam is extremely short, it should be *chosen* for moment and *checked* for shear; that is, the shear stress f_v should be less than the allowable value F_v. The shear stress as computed by simple bending theory at any location in the beam web is given by

$$f_v = \frac{VQ}{It} \qquad (3.8)$$

where

> V = total resultant shear force on cross section
>
> Q = static moment, taken about the neutral axis, of that portion of the beam area beyond the point at which the shear stress is to be calculated
>
> t = web thickness where the stress is computed

The designer may use Eq. (3.8) for certain shear-dependent details, such as the welds connecting the web and flange of built-up beams, or the webs of unsymmetrical sections. But simpler expressions for the web shear stress are specified for the usual beam design situation. For a W or C beam section, in simple bending, the shear stress is approximated by dividing the resultant shear at any location by the product of the web thickness times the full depth of the beam. In the design of built-up girders the area is based on the depth of the girder web plate between flange plates. Figure 3.7 illustrates the three alternatives just discussed.

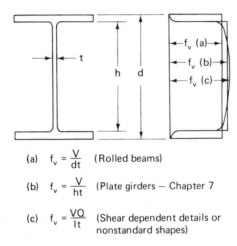

(a) $f_v = \dfrac{V}{dt}$ (Rolled beams)

(b) $f_v = \dfrac{V}{ht}$ (Plate girders — Chapter 7

(c) $f_v = \dfrac{VQ}{It}$ (Shear dependent details or nonstandard shapes)

Fig. 3.7 Three alternatives for the estimate of web shear stress due to bending.

3.3   INELASTIC BEHAVIOR OF STEEL BEAMS

If *static* load capacity is the primary basis for the choice of a beam, the selection may be made on the basis of maximum strength without recourse to stress calculations. This is standard procedure in *plastic design*, which is based both on the plastic moment capacity of the individual beam and on the ultimate strength of the complete continuous-frame structure. Moreover, the considerable variation in allowable stresses in *elastic design* can be justified, as will be shown, on the basis of maximum-strength criteria.

Referring back to Fig. 1.2, let it be assumed that the stress–strain diagram for structural steel consists simply of the two straight-line portions out to the initiation of strain hardening and labeled "elastic range" and "plastic range," respectively. In this case, the maximum stress due to bending in a steel beam would not rise above the yield stress (F_y). Figure 3.8(b) shows the stress distribution assumed in allowable stress design. If the bending moment is

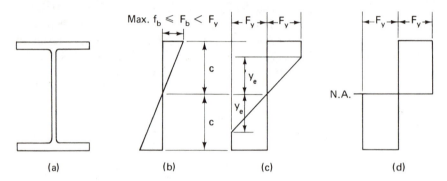

Fig. 3.8 Distribution of normal stress due to bending in elastic and inelastic ranges.

increased above the amount that causes the stress to be equal to F_b, until the stress f_b just equals F_y, the stress diagram will continue to have the elastic, linear distribution shown in Fig. 3.8(b), and the bending moment will have reached M_y, the yield moment. Above the yield moment, the stress distribution will be as shown in Fig. 3.8(c), finally approaching in the limit the rectangular shape shown in Fig. 3.8(d), which corresponds to the plastic moment, M_p, the maximum attainable if no strain hardening were to occur.

The inelastic behavior of a steel beam is best illustrated by a plot of bending moment versus curvature. Above M_y the curvature is no longer linearly related to moment by the elastic relationship of Eq. (3.3). At M_y the curvature $\phi_y = F_y/Ec = M_y/EI$. Above M_y, at moments less than M_p, and using the notation shown in Fig. 3.8(c), the curvature is equal to F_y/Ey_e. M versus ϕ curves for circular, rectangular, and W beam shapes are shown in Fig. 3.9. In these plots each shape has the same section modulus and the same M_y. The ratio M_p/M_y is called the *shape factor*, a measure of the increase in plastic moment strength in comparison with the yield moment. The circular section is the most inefficient from an elastic point of view, although it absorbs more energy than the other shapes before reaching the yield moment. The average shape factor for the W shape is only 1.14, but this is a least-weight shape designed to be efficient in the elastic range of behavior. To ensure an

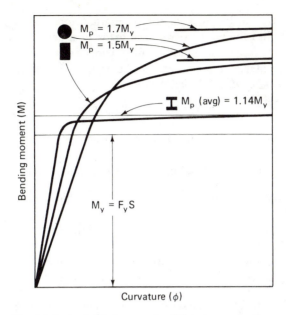

Fig. 3.9 Inelastic M-ϕ curves for different cross sections.

approach toward M_p without local flange buckling or lateral buckling, the wide-flange shape must be "compact," as defined by AISCS, Sec. 1.5.1.4.1. If not compact, design is to be by the allowable-stress procedure, that is, based on M_y and not M_p.

The AISCS allowable-stress design procedure permits bending moments tabulated below as $M_{\text{allowable}}$ for the four different shapes just discussed.

	$M_{\text{allowable}}$	M_{max}	$M_{\text{max}}/M_{\text{allowable}}$
Solid circular section	$0.75F_yS$†	$M_p = 1.7F_yS$	2.27
Rectangular section	$0.75F_yS$†	$M_p = 1.5F_yS$	2.00
Compact W shape	$0.66F_yS$†	$M_p = 1.14F_yS$ (av.)	1.73
Noncompact W shape	$0.60F_yS$	$M_y = F_yS$	1.67

†Not applicable to ASTM A514 steel.

The maximum moment is listed as the plastic moment for the circular, rectangular, and compact W shapes; but for the noncompact W shape there is no assurance of attaining and holding any moment greater than that for which the maximum stress first reaches the yield point. The strength load factors are tabulated in the last column. The load factors illustrate how adjustments in allowable stress have been used to at least partially compensate for wide variations in the plastic ultimate strength.

To design a beam by the maximum-strength, or plastic-design procedure, one first multiplies the required working loads by a design load factor, and determines the required maximum moment capacity at failure. This must be exceeded by the plastic moment (M_p) of the beam to be selected. As an index of M_p, the *plastic modulus*, Z, is introduced:

$$Z = \frac{M_p}{F_y} = S \times \text{(shape factor)} \tag{3.9}$$

The plastic modulus, Z, is tabulated for compact rolled W and S shapes in the AISCM.

In summary, then, beam selection by plastic design involves:

1. Determination of maximum moment at required ultimate capacity load, or working load times a load factor.
2. Determination of required Z, which is moment determined in step 1 divided by yield stress F_y.
3. Selection of beam with Z equal to or greater than that required.
4. Checking the compactness of section and shear strength.

In the case of simple beams the selection will usually not be different from that obtained by allowable-stress design.

The real advantage of plastic design results from application to continuous beams or frames, to be covered in Chapter 8.

3.4 ALLOWABLE STRESSES FOR ELASTIC DESIGN

In the allowable-stress design procedure, steel beams are selected so that the maximum normal and shear stress components due to bending do not exceed the AISCS allowable values for tension which are specified in terms of the yield strength F_y. The allowable stresses in kips per square inch (ksi) are also tabulated for the various available yield points in the AISCS, Appendix A, Table 1. In some cases the listed values have been rounded off. For example, when the allowable stress is $0.66F_y$, it would be $0.66 \times 36 = 23.76$ ksi, but the allowable value as tabulated in the Appendix has been rounded off to 24 ksi. For convenience, the rounded-off values, as specified, have been listed herein for most steels in Table 2.1.

Compact Sections

AISCS, Sec. 1.5.1.4.1, states that the allowable stress on extreme fibers of *compact* hot-rolled sections (except A514 steel) symmetrical about and loaded in the plane of their minor axis shall be

$$F_b = 0.66F_y$$

The "basic" allowable stress in tension or compression is $0.6F_y$. The compact sections, for which a stress of $0.66F_y$ is permitted, are required to have relatively thicker flanges and webs than the noncompact and must have added safeguards against local and lateral buckling. The resulting bonus of 10% in the allowable stress is for sections which, as discussed in Section 3.3, will develop their plastic moment strength, which is, on the average, 14% greater than the yield moment strength. It is this difference between yield moment and plastic moment that establishes the logic of the stress increase.

The specific requirements are, by AISCS:

1. The flanges shall be continuously connected to the web. (A built-up section with intermittent welds would not qualify.)

2. For the unstiffened projecting elements of the compression flange, the width/thickness ratio shall be based on half the full flange width (AISCS, Sec. 1.9.1.1), and $b_f/2t_f$ shall not exceed $65.0/\sqrt{F_y}$, where b_f is the width of the compression flange, and t_f is the thickness of the compression flange.

F_y	36	40	42	46	50	60	65
$\dfrac{65.0}{\sqrt{F_y}}$	10.8	10.3	10.0	9.6	9.2	8.4	8.1

The "unstiffened" projecting element refers to the usual situation in standard hot-rolled sections, such as the W shape. Stiffened elements include flanges with lipped edges, sometimes used in cold-formed shapes. The flange of a box beam flanked by two webs would also be considered a stiffened element.

3. For the stiffened elements of the compression flange, the width/thickness ratio (b/t_f) as defined in AISCS, Sec. 1.9.2.1, shall not exceed

$$\frac{190}{\sqrt{F_y}} \qquad (b = \text{actual width of stiffened plate segment})$$

F_y	36	40	42	46	50	60	65
$\dfrac{190}{\sqrt{F_y}}$	31.7	30.0	29.3	28.0	26.9	24.5	23.6

4. The depth/thickness ratio of the web shall not exceed

$$\frac{640}{\sqrt{F_y}}$$

F_y	36	40	42	46	50	60	65
$\dfrac{640}{\sqrt{F_y}}$	106.7	101.2	98.8	94.4	90.5	82.6	79.4

It should be noted that the foregoing rule for maximum d/t results from the AISCS, Eq. (1.5-4a) when $f_a = 0$. The case for which $f_a \neq 0$ will be considered in Chapter 5. The foregoing limitation on d/t is intended as a safeguard against premature web buckling in a beam or girder without transverse stiffeners.

5. The compression flange for members other than circular or box members shall be supported laterally, at intervals l_b, not to exceed either of the following limits:

$$\frac{l_b}{b_f} \leq \frac{76.0}{\sqrt{F_y}}, \qquad \frac{l_b d}{A_f} \leq \frac{20,000}{F_y}$$

F_y	36	40	42	46	50	60	65
$\dfrac{76.0}{\sqrt{F_y}}$	12.7	12.0	11.7	11.2	10.7	9.8	9.4
$\dfrac{20,000}{F_y}$	556	500	476	435	400	333	308

6. The compression flange of a box member, whose depth is not more than six times the width ($d \leq 6b$) and whose flange thickness is not more than two times the web thickness ($t_f \leq 2t_w$), shall be supported laterally, at intervals l_b, not to exceed the larger value of the following two limits:

$$\frac{1200 (1.625 + M_1/M_2)b}{F_y}$$

$$\frac{1200b}{F_y}$$

where M_1 is the smaller and M_2 the larger bending moment at the ends of the unbraced length, and where M_1/M_2, the end-moment ratio, is positive when the beam segment is in reverse (double) curvature bending and negative when it is in single curvature bending.

7. For circular sections, the diameter/thickness (D/t) ratio shall not exceed $3300/F_y$.

F_y	36	40	42	46	50	60	65
$\dfrac{1200}{F_y}$	33.3	30.0	28.6	26.1	24.0	20.0	18.5
$\dfrac{3300}{F_y}$	91.7	82.5	78.6	71.7	66.0	55.0	50.8

Transition Between Compact and Noncompact Members

If a beam with an unstiffened flange meets all the requirements for a compact section except that $b_f/2t_f$ exceeds $65.0/\sqrt{F_y}$, AISCS Sec. 1.5.1.4.2 provides for a transition in allowable stress between the values of $0.66F_y$ and $0.60F_y$.

This does not apply to hybrid girders or to members of A514 steel, nor is the transition formula applied if it indicates an allowable stress less than $0.60F_y$, for which $b_f/2t_f$ would exceed $95.0/\sqrt{F_y}$. The transition formula [Eq. (1.5-5a)] is

$$F_b = F_y\left(0.79 - 0.002\frac{b_f}{2t_f}\sqrt{F_y}\right)$$

Width/Thickness Limitation for Noncompact Members

To be fully effective at the allowable stress of $0.6F_y$, the AISCS (Sec. 1.9.1.2) requires that the width/thickness ratio of an unstiffened half flange ($b_f/2t_f$) be no greater than $95.0/\sqrt{F_y}$, as tabulated below for the various yield points.

F_y	36	40	42	46	50	60	65	90	100
$\dfrac{95.0}{\sqrt{F_y}}$	15.8	15.0	14.7	14.0	13.4	12.3	11.8	10.0	9.5

Members with outstanding unstiffened compression elements exceeding the values above must be designed for average allowable stresses less than $0.6F_y$, as covered in Appendix C of the AISCS.

Solid Round, Square, or Rectangular Sections

As discussed in Section 3.3, rectangular and round sections have plastic strength shape factors of 1.5 and about 1.7, respectively, with corresponding increases in ultimate moment capacity in relation to bending moment at initial yield. Thus, if designed by allowable-stress procedures, it is reasonable to permit an increase in allowable stress. W and S sections bent about the weak axis, provided that the width/thickness ratio requirements for compact sections are met, are essentially similar in bending behavior to the solid rectangular section. Thus, excepting A514 steel, AISCS, Sec. 1.5.1.4.3, allows a stress in bending of $F_b = 0.75F_y$, provided, in the case of W and I sections, that the provisions of Sec. 1.5.1.4.1 (1) and (2) are met (see Table 2.1).

When $b_f/2t_f$ is between $65.0/\sqrt{F_y}$ and $95.0/\sqrt{F_y}$, an otherwise compact shape, bent about the weak axis, may be designed for an allowable stress between $0.75F_y$ and $0.60F_y$, in accord with transition equation (1.5-5b):

$$F_b = F_y\left(1.075 - 0.005\frac{b_f}{2t_f}\sqrt{F_y}\right)$$

Box and Circular Tube Sections

Box sections are especially recommended for design situations involving incomplete lateral support. Failure by lateral-torsional buckling involves twisting in combination with lateral bending about the weak axis. Box sections are greatly superior to I or W (open) sections in both of these characteristics. Standard structural (box) tubes, as shown in Fig. 3.10(a), are cataloged in the AISCM. Box beams also may be built up as a welded plate assemblage [Fig. 3.10(b)].

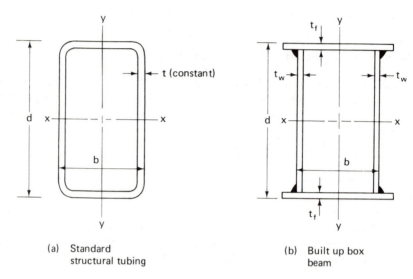

(a)   Standard structural tubing

(b)   Built up box beam

Fig. 3.10 Box beams.

The AISCS (Sec. 1.5.1.4.1) permits the allowable stress for compact sections to be applied to box members, provided that requirements (1) through (6) are met.

Box sections with a depth/width (d/b) ratio less than 6 that do not meet the requirements for compact sections (AISCS, Sec. 1.5.1.4.1), but do conform to the requirements of Sec. 1.9.2.2 ($b/t \leq 238/\sqrt{F_y}$), should be designed for the basic allowable stress of $0.6F_y$ (AISCS, Sec. 1.5.1.4.4).

F_y	36	40	42	46	50	60	65	90	100
$\dfrac{238}{\sqrt{F_y}}$	39.7	37.6	36.7	35.1	33.7	30.7	29.5	25.1	23.8

Box sections that do not meet the foregoing requirements for d/b and b/t should be designed for an allowable stress less than $0.6F_y$, as determined by a special lateral-torsional buckling analysis.[†]

Circular tubes meeting the requirement Sec. 1.5.1.4.1(7) made of other than A514 steel may be designed for an AISCS allowable stress of $0.66F_y$,

[†]Reference should be made to Chapter 10 for design examples utilizing box beams.

but if the diameter/thickness ratio exceeds $3300/F_y$, the allowable stress is less than $0.60F_y$ and AISCS Appendix C should be referred to.

Miscellaneous Sections

In general, for any steel shape not explicitly cited, the maximum allowable tension in flexural members is $0.6F_y$, as per AISCS, Sec. 1.5.1.4.5. The accepted rounded-off values are listed in Table 2.1, herein, and in AISCS, Appendix A, Table 1.

W and C Shapes with Incomplete Lateral Support

W shapes are designed to be highly efficient when loaded in the plane of the web and supported laterally. When the lateral support is insufficient, the designer must decide as to which of several alternatives will afford the greatest economy. These include:

1. Rearrange the framing to provide better lateral support.
2. Change to a box section, at higher cost per pound of material.
3. Select a W or C shape at an allowable stress reduced below the basic allowable of $0.6F_y$.

If the first choice is either not feasible or too costly, the third choice will probably prove the economical one if the reduction in allowable stress is relatively small. The stress-reduction alternative is covered by AISCS, Sec. 1.5.1.4.5, which provides two formulas, the first of which is in two parts, (1.5-6a) and (1.5-6b). The allowable stress is the *larger* obtained by trying the two formulas, but in no case is the allowable stress to be greater than the basic value of $0.6F_y$. In manual computation, the second formula is the easiest to compute; hence it is recommended that the designer initially check Formula (1.5-7), and if the result exceeds $0.6F_y$, there is no need to go to Formulas (1.5-6). The AISCS Commentary should be referred to for justification for the use of two formulas.

When a beam fails by lateral-torsional buckling, it bends (buckles) about its *weak* axis, even though loaded normally in the strong plane so as to bend about its *strong* axis—which, indeed, it does, up to the critical load at which it buckles. When the beam buckles laterally about its weak axis, the loads also induce a torsional moment in the beam. The torsional resistance† of a W, S, or C section is made up of two parts: (1) the minimal torsion resistance that would be obtained under uniform torsion alone, plus (2) torsional resistance due to coupled bending of the flanges, inducing shears in the flanges that create a torsion couple. The torsion that accompanies lateral buckling is

---

†The torsional resistance of W shapes is explained in greater detail in Chapter 10.

always nonuniform. Thus the resistance to lateral-torsional buckling of a W beam consists of three parts:

1. Lateral bending about the weak axis.
2. Uniform torsion resistance (St. Venant torsion).
3. Nonuniform torsion resistance (warping torsion).

In the interest of obtaining a simple procedure, AISCS Formulas (1.5-6) neglect the uniform torsion contribution, whereas Formula (1.5-7) neglects the nonuniform torsion contribution. Comparing the two contributions to torsional resistance, uniform torsion is relatively the smaller in thin-walled, deep sections, such as plate girders. Nonuniform, or warping torsion, is the smaller in thick-walled members of relatively long span. Section 1.5.1.4.5(2) of AISCS reads as follows:

2. Compression:
   a. For members meeting the requirements of Sec. 1.9.1.2, having an axis of symmetry in, and loaded in, the plane of their web, and compression on extreme fibers of channels bent about their major axis:
   The larger value computed by Formulas (1.5-6a) or (1.5-6b) and (1.5-7), as applicable* (unless a higher value can be justified on the basis of a more precise analysis†), but not more than $0.60F_y$‡

When $\sqrt{\dfrac{102 \times 10^3 C_b}{F_y}} \le \dfrac{l}{r_T} \le \sqrt{\dfrac{510 \times 10^3 C_b}{F_y}}$:

$$F_b = \left[\frac{2}{3} - \frac{F_y\,(l/r_T)^2}{1530 \times 10^3 C_b} \right] F_y \qquad (1.5\text{-}6a)$$

When $\dfrac{l}{r_T} \ge \sqrt{\dfrac{510 \times 10^3 C_b}{F_y}}$:

$$F_b = \frac{170 \times 10^3 C_b}{(l/r_T)^2} \qquad (1.5\text{-}6b)$$

Or, when the compression flange is solid and approximately rectangular in cross section and its area is not less than that of the tension flange:

$$F_b = \frac{12 \times 10^3 C_b}{ld/A_f} \qquad (1.5\text{-}7)$$

In the foregoing,

l = distance between cross sections braced against twist or lateral displacement of the compression flange, inches. For canti-

---

*Only Formula (1.5-7) applicable to channels.

†See Commentary Sec. 1.5.1.4.5 for alternate procedures.

‡See Sec. 1.10 for further limitations in plate girder flange stress.

levers braced against twist only at the support, l may conservatively be taken as the actual length.

r_T = radius of gyration of a section comprising the compression flange plus $\frac{1}{3}$ of the compression web area, taken about an axis in the plane of the web, inches

A_f = area of the compression flange, square inches

C_b = $1.75 + 1.05 (M_1/M_2) + 0.3 (M_1/M_2)^2$, but not more than 2.3,§ where M_1 is the smaller and M_2 the larger bending moment at the ends of the unbraced length, taken about the strong axis of the member, and where M_1/M_2, the ratio of end moments, is positive when M_1 and M_2 have the same sign (reverse curvature bending) and negative when they are of opposite signs (single curvature bending). When the bending moment at any point within an unbraced length is larger than that at both ends of this length, the value of C_b shall be taken as unity. When computing F_{bx} and F_{by} to be used in Formula (1.6-1a), C_b may be computed by the formula given above for frames subject to joint translation, and shall be taken as unity for frames braced against joint translation. C_b may conservatively be taken as unity for cantilever beams.¶

For hybrid plate girders, F_y for Formulas (1.5-6a) and (1.5-6b) is the yield stress of the compression flange. Formula (1.5-7) shall not apply to hybrid girders.

b. For members meeting the requirements of Sec. 1.9.1.2, but not included in subparagraph 2a of this Section:

$$F_b = 0.60F_y$$

provided that sections bent about their major axis are braced laterally in the region of compression stress at intervals not exceeding $76b_f/\sqrt{F_y}$.

AISCS Formulas (1.5-6) and (1.5-7) may be termed "semirational" in that they are simplifications of theoretical formulas for critical stress. The term C_b is intended to account for situations wherein the bending moment varies over the unbraced length. For uniform moment, $M_1 = -M_2$ and C_b is then unity, which is its minimum value. Values greater than unity should be restricted to situations for which the variation in bending moment within the unbraced length is linear, or nearly so, which will be the case if lateral bracing coincides with points of concentrated load application.

The AISCM provides listings of both r_T and d/A_f for rolled members

§C_b can be conservatively taken as unity. For smaller values see Appendix A, Table 7.

¶For the use of larger C_b values, see Structural Stability Research Council, *Guide to Stability Design Criteria for Metal Structures*, 3rd edition, p. 135.

used as beams, thus facilitating the numerical application of Formulas (1.5-6) and (1.5-7).

Allowable Stress in Shear

Except in the case of very short spans, beams are usually selected on the basis of allowable stress in bending, then checked for the shear stress for which (AISCS, Sec. 1.5.1.2) the allowable is $0.4F_y$. Reference is made to AISCS, Sec. 1.10, for required reduction in this allowable value for very thin webs. The use of transverse stiffeners to increase the permissible shear stress in thin webs, also covered by Sec. 1.10, will be treated in Chapter 7. Formula (1.10-1), with $F_v = 0.4F_y$, requires that $C_v = 1.156$. As a/h becomes very large, the case with no transverse stiffeners, $k = 5.34$. These values of C_v and k then provide the following limit on h/t for which the full allowable shear stress $F_v = 0.4F_y$ may be used:

$$\frac{h}{t} \leq \frac{380}{\sqrt{F_y}}$$

Note that h is the clear depth between flanges, not the full depth d of the beam. For the various yield stresses:

F_y	36	40	42	46	50	60	65	90	100
$0.4F_y$	14.5	16.0	16.8	18.4	20.0	24.0	26.0	36.0	40.0
$\dfrac{380}{\sqrt{F_y}}$	63.3	60.1	58.6	56.0	53.7	49.1	47.1	40.0	38.0

The foregoing tabulation does not rule out the use of beam sections with greater h/t values, provided that a reduced allowable shear stress (less than $0.4F_v$) is specified according to the rules for thin-web girder design.

3.5   LATERAL SUPPORT REQUIREMENTS

Most beams are designed by simple bending theory with the assumption of full lateral support and without any required reduction in allowable stress due to bending. Any beam with the compression flange attached securely to a floor or roof system that provides continuous or nearly continuous support meets these requirements.

Some conditions for which lateral support may be less than adequate include the following:

1. No positive connection between the beam and the load system that it supports, particularly if loads are vibratory or involve impact.
2. The lateral support system is of a removal type.
3. Lateral support system frames into a parallel system of two or more similarly loaded beams without positive anchorage. This possibility is illustrated in Fig. 3.11, which shows three alternative framing systems in plan view. In Fig. 3.11(a) the lateral support is inadequate for the reason cited. In Fig. 3.11(b) and (c) it is adequate, because of adjacent beams with wall anchorage or by virtue of K-bracing that limits motion, respectively.

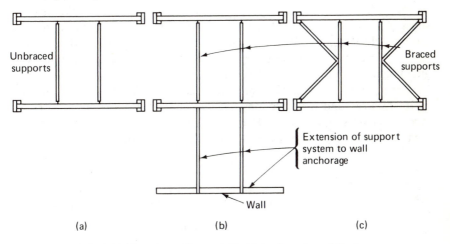

(a)   (b)   (c)

Fig. 3.11 Plan view of beams with (a) inadequate and (b, c) adequate lateral support systems.

3.6 BEAM DEFLECTION LIMITATIONS

When plastered ceilings are attached to beams, the AISCS, Sec. 1.13.1, limits the maximum live-load deflection to not more than $\frac{1}{360}$ of the span. This is an empirical relationship, based primarily on past experience, and intended to prevent or limit unsightly cracking in plastered surfaces.

Limits on deflection may be imposed indirectly by limitations on beam depth/span ratios. The interrelationship among span, stress, and deflection may be illustrated as follows. Consider the case of the uniformly loaded simple span; the deflection is

$$\Delta = \frac{5wl^4}{384EI} = \frac{5}{48}\frac{wl^2}{8}\frac{l^2}{EI} = \frac{5M_{max}l^2}{48EI} \tag{3.10}$$

The maximum stress due to bending is

$$f_b = \frac{M_{max}d}{2I}$$

which gives the following:

$$\frac{l}{d} = \frac{139,200}{f_b(l/\Delta)} \qquad \text{for } E = 29,000 \text{ ksi} \qquad (3.11)$$

The AISCS Commentary (Sec. 1.13.1) suggests that the depth/span ratio be not less than $F_y/800$ for fully stressed beams and girders in floors, with the provision that this ratio may be decreased in the same ratio as the bending stress is decreased from that for a fully stressed beam. In the case of roof purlins, fully stressed, it is recommended that the depth/span ratio be not less than $F_y/1000$, except in the case of flat roofs, where special provisions to guard against "ponding" may be required.

For $f_b = F_b = 0.6F_y$, the foregoing recommendations of $F_y/800$ and $F_y/1000$ correspond to deflections of $l/290$ and $l/232$, respectively, for uniformly loaded simple beams. For the available yield points, these recommendations correspond to the following listed values of l/d:

F_y	36	40	42	46	50	60	65	90	100
$\dfrac{800}{F_y}$	22.2	20.0	19.0	17.4	16.0	13.3	12.3	8.9	8.0
$\dfrac{1000}{F_y}$	27.8	25.0	23.8	21.7	20.0	16.7	15.4	11.1	10.0

If water on a flat roof accumulates at a rate more rapid than it runs off, additional roof load develops due to the water that is "ponded" as a result of roof deflection, thus further aggravating the imbalance between rainfall and runoff rates. Failure may result. The problem is treated in the AISCS Sec. 1.13.3 and in the AISCS Commentary. In flat floor or roof systems of long span, the supporting girders should be cambered to minimize the ponding hazard if it exists, and to avoid any visually perceptible sag.

3.7 BEAMS UNDER REPEATED LOAD

Design of beams for repeated load by the allowable-stress-range procedure is essentially the same as for tension members, which has been covered in Chapter 2. Typical beam category designations are illustrated in Appendix B of the AISCS.

In beam design it may be possible to effect economy by changing the arrangement of certain welded details so as to place the beam in the most favorable repeated load stress category (see Example 3.4).

3.8 BIAXIAL BENDING OF BEAMS

The W beam, or a modification thereof as shown in Fig. 3.12(b), is frequently used in design situations for which components of bending moment occur simultaneously about both the x–x and y–y axes. If the allowable stress about

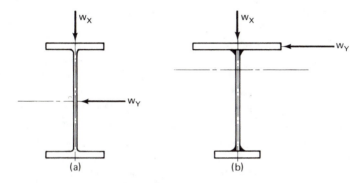

Fig. 3.12 Beams under biaxial load.

the two axes were the same, the design could be based simply on the maximum stress calculated by superposing the stresses caused by bending about each of the two principal axes:

$$f_b = \pm \frac{M_x}{S_x} \pm \frac{M_y}{S_y} \tag{3.12}$$

If the component of load in the weak plane is comparable in magnitude to that in the strong plane, a box section may be the preferred solution. If the lateral loads are relatively small, the W shape may be suitable. Because of the difference in allowable stresses about the two axes, an interaction formula—AISCS, Formula (1.6-2), with f_a set equal to zero—should be used. As modified, this becomes

$$\frac{f_{bx}}{F_{bx}} + \frac{f_{by}}{F_{by}} = 1.0 \tag{3.13}$$

Since F_{bx} and F_{by} depend on the selection of W shape, one cannot make a direct estimate of the required section modulus. However, an approximate approach may be made by introducing the following tentative assumptions:

$$F_{bx} = 0.60F_y \qquad F_{by} = 0.75F_y \qquad S_y = \frac{S_x}{C_n}$$

With these substitutions a formula for the preliminary selection on the basis of section modulus S_x is obtained:

$$S_x = \frac{M_x}{0.60F_y} + \frac{C_n M_y}{0.75F_y} \tag{3.14}$$

The coefficient C_n may be chosen on the basis of the following guidelines. If it is probable that the W section is in the 27- to 36-in. depth range, set $C_n = 7$, which is an approximation for the wider of the two W widths in this depth range. The wider sections will be most effective in resisting lateral loads and will also offer the greatest values of F_{bx}, which will be subject to reduction according to AISCS, Sec. 1.5.1.4.5. In the 16- to 24-in. depth range, there are four different sets of widths available and $C_n = 5$ could be used for preliminary trial selections. For depths of 14 in. and less, W column shapes are available as wide or wider than they are deep. For these, C_n ranges between 2.0 and 2.5 for most sections. Of course, if the lateral loads are only a very small proportion of the vertical loads, it will be more economical to use a section having C_n in the range 5 to 7.

If the lateral load is applied at or near the top flange, the use of an unsymmetrical section built up by welding three plates, as shown in Fig. 3.12(b), may be the most economical solution. The complex consideration of the torsion that is introduced may be avoided by calculating the stress in the top flange, due to lateral load, under the assumption that the top flange alone carries *all* the lateral load. This simplification of the analysis problem provides a safe design procedure. In applying the modified AISCS interaction formula (1.6-2), F_{bx} may be calculated by AISCS, Formula (1.5-6), with f_{bx} as determined for the compression flange. In the tension flange it may be assumed that the stress is unaffected by lateral load, but f_{bx} will be greater than in the compression flange.

Alternative approaches to the biaxial beam design problem are presented in Section 3.12.

3.9 LOAD AND SUPPORT DETAILS

Individual beams may carry concentrated loads and must be supported at or near their ends. Concentrated loads may be introduced directly into a beam web by means of a riveted, bolted, or welded beam web connection, and the loaded beam may in turn transmit its end reactions to columns or girders by means of similar connections. Beam web connections of the type just described are covered in Chapter 6.

When the end of a beam rests on masonry, it requires a bearing support, such as is shown in Fig. 3.13. Consideration must be given to the local compression stress in the web, just above the bearing block, and to the required

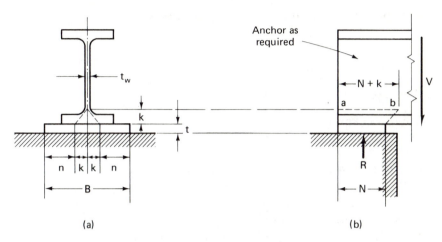

Fig. 3.13 Bearing supports at end of a beam.

thickness of bearing plate to spread the load to the masonry at a permissible pressure, F_p. The local concentration of compression in the beam web is assumed uniformly distributed over the distance $(N + k)$, as shown in Fig. 3.13. k is tabulated in the W shape detailing information tables in the AISCM, and is the distance from the flange face to the termination of the flange to web fillet. The compression stress along this line, equal to $R/(N + k)t_w$, must be kept below $0.75F_y$ (AISCS, Sec. 1.10.10). Note that the pressure is assumed to spread out from the edge of the bearing block at a 45° angle. If a support occurs away from the end of a beam, or if a local load is introduced at the top of a beam through a bearing block, the situation is similar, except the spread of load proceeds from each end of the bearing block and the compression stress is assumed to be $R/(N + 2k)t_w$.

Bearing plates must be thick enough to spread the reactions or concentrated loads to the masonry at allowable pressures, F_p, as specified in AISCS, Sec. 1.5.5. The required bearing plate thickness is determined by considering the bearing plate as a simple cantilever beam of length n, as shown in Fig. 3.13(a). The beam carries an upward load, assumed uniform, resulting from the masonry pressure. It is assumed that the plate will distribute the load satisfactorily if the stress due to bending in the cantilever beam is kept below $0.75F_y$ (AISCS, Sec. 1.5.1.4.3). A formula for bearing plate thickness is readily derived. Assume a unit width of cantilever beam:

$$M_{\max} = (f_p)(1)(n)\frac{n}{2} = \frac{f_p n^2}{2}$$

$$S = \frac{(1)(t^2)}{6}$$

$$f_{\max} = 0.75F_y = \frac{M}{S} = \frac{3f_p n^2}{t^2}$$

Solving for t, the required bearing plate thickness is

$$t_{\text{reqd}} = n\sqrt{\frac{4f_p}{F_y}} \qquad (3.15)$$

For A36 steel this simplifies further to

$$t_{\text{reqd}} = \frac{n}{3}\sqrt{f_p} \qquad (3.16)$$

where f_p is in kips per square inch.

3.10 ALLOWABLE LOAD TABLES FOR BEAMS

The AISCM provides tables to permit direct selection of beams under uniform load for yield stresses of either 36 or 50 ksi. Information is also supplied regarding maximum unbraced lengths. In addition, design coefficients are provided to expedite calculation of allowable loads for beams having yield points other than 36 or 50 ksi. If the load and/or support conditions are other than a simple beam under uniform load, the allowable load tables may be used in many instances by conversion of the different load situation to an "equivalent (uniform) tabular load." Conversion factors are given in the AISCM for a variety of load and support conditions.

Deflection coefficients for beams in terms of span and maximum stress due to bending permit rapid calculation of center deflection of uniformly loaded beams. Modifying factors permit good approximations for other load conditions.

3.11 FLOWCHARTS FOR STEEL BEAM DESIGN

The flowcharts presented on the following pages are based primarily on the AISCS and also include, as noted, several recommendations for design by the authors.

The flowcharts include:

Flowchart 3.1: Allowable stress due to flexure, F_b. It will be called for during the use of both the steel beam and column flow charts, the latter in Chapters 4 and 5.

Flowchart 3.2: Steel beam selection. This chart includes W, S, C, circular, and box-shaped beams. Biaxial bending is included insofar as the section has at least one axis of symmetry. Insofar as channels are covered, lateral supports would need to be provided at support and load locations.

Flowchart 3.1

(1) Allowable Bending Stress About Major Axis (AISCS)

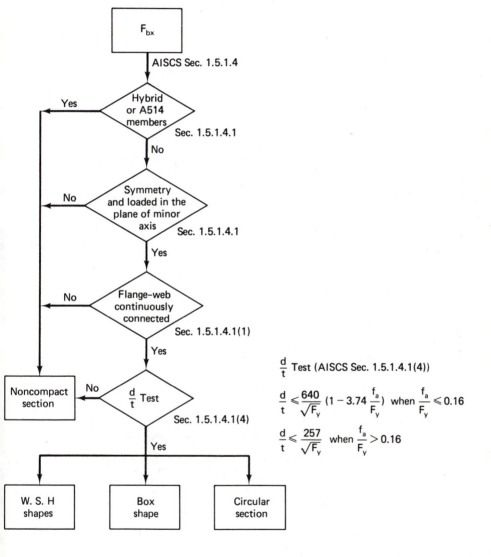

$\dfrac{d}{t}$ Test (AISCS Sec. 1.5.1.4.1(4))

$$\frac{d}{t} \leqslant \frac{640}{\sqrt{F_y}} \, (1 - 3.74 \, \frac{f_a}{F_y}) \quad \text{when} \quad \frac{f_a}{F_y} \leqslant 0.16$$

$$\frac{d}{t} \leqslant \frac{257}{\sqrt{F_y}} \quad \text{when} \quad \frac{f_a}{F_y} > 0.16$$

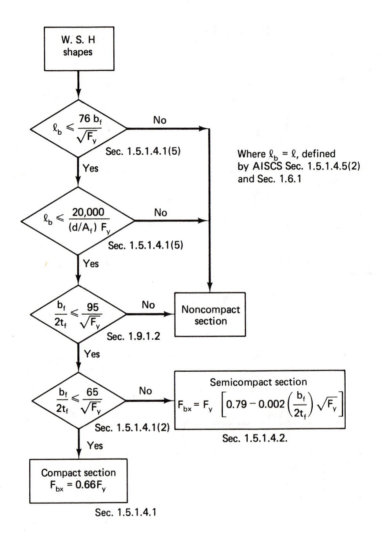

W. S. H shapes

$\ell_b \leq \dfrac{76\,b_f}{\sqrt{F_y}}$ No

Sec. 1.5.1.4.1(5)

Yes

Where $\ell_b = \ell$, defined by AISCS Sec. 1.5.1.4.5(2) and Sec. 1.6.1

$\ell_b \leq \dfrac{20,000}{(d/A_f)\,F_y}$ No

Sec. 1.5.1.4.1(5)

Yes

$\dfrac{b_f}{2t_f} \leq \dfrac{95}{\sqrt{F_y}}$ No

Sec. 1.9.1.2

Noncompact section

Yes

$\dfrac{b_f}{2t_f} \leq \dfrac{65}{\sqrt{F_y}}$ No

Sec. 1.5.1.4.1(2)

Semicompact section

$F_{bx} = F_y \left[0.79 - 0.002 \left(\dfrac{b_f}{2t_f} \right) \sqrt{F_y} \right]$

Sec. 1.5.1.4.2.

Yes

Compact section
$F_{bx} = 0.66 F_y$

Sec. 1.5.1.4.1

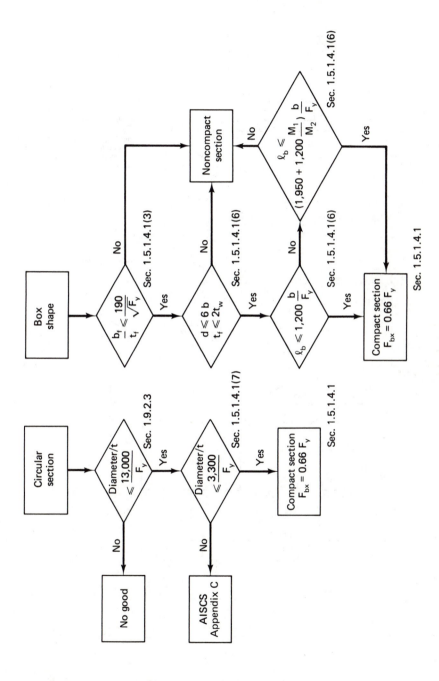

Box shape

$\dfrac{b_f}{t_f} \le \dfrac{190}{\sqrt{F_y}}$ Sec. 1.5.1.4.1(3)

No → **Noncompact section**

Yes ↓

$\dfrac{d}{t_f} \le \dfrac{6\,b}{2t_w}$ Sec. 1.5.1.4.1(6)

No → **Noncompact section**

Yes ↓

$\ell_b \le 1,200\,\dfrac{b}{F_y}$ Sec. 1.5.1.4.1(6)

No ↓

$\ell_b \le \left(1,950 + 1,200\,\dfrac{M_1}{M_2}\right)\dfrac{b}{F_y}$ Sec. 1.5.1.4.1(6)

No → **Noncompact section**

Yes →

Yes → **Compact section** $F_{bx} = 0.66\,F_y$ Sec. 1.5.1.4.1

Circular section

$\text{Diameter}/t \le \dfrac{13,000}{F_y}$ Sec. 1.9.2.3

No → **No good**

Yes ↓

$\text{Diameter}/t \le \dfrac{3,300}{F_y}$ Sec. 1.5.1.4.1(7)

No → **AISCS Appendix C**

Yes → **Compact section** $F_{bx} = 0.66\,F_y$ Sec. 1.5.1.4.1

73

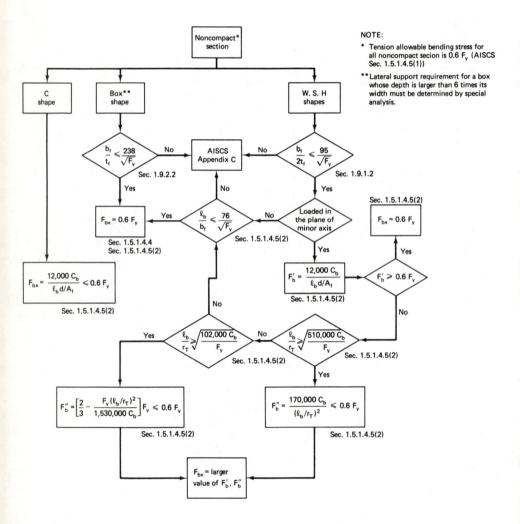

NOTE:

* Tension allowable bending stress for all noncompact secion is 0.6 F_y (AISCS Sec. 1.5.1.4.5(1))

** Lateral support requirement for a box whose depth is larger than 6 times its width must be determined by special analysis.

Noncompact* section

C shape

Box** shape

W. S. H shapes

$\dfrac{b_f}{t_f} \leqslant \dfrac{238}{\sqrt{F_y}}$ No

Sec. 1.9.2.2

AISCS Appendix C

No

$\dfrac{b_f}{2t_f} \leqslant \dfrac{95}{\sqrt{F_y}}$

Sec. 1.9.1.2

Yes

Yes

$F_{bx} = 0.6 \, F_y$

Sec. 1.5.1.4.4
Sec. 1.5.1.4.5(2)

Yes

$\dfrac{\ell_b}{b_f} \leqslant \dfrac{76}{\sqrt{F_y}}$ No

Sec. 1.5.1.4.5(2)

No

Loaded in the plane of minor axis

Yes

Sec. 1.5.1.4.5(2)

$F_{bx} = 0.6 \, F_y$

Yes

$F_{bx} = \dfrac{12,000 \, C_b}{\ell_b d/A_f} \leqslant 0.6 \, F_y$

Sec. 1.5.1.4.5(2)

$F_b' = \dfrac{12,000 \, C_b}{\ell_b d/A_f}$

Sec. 1.5.1.4.5(2)

$F_b' \geqslant 0.6 \, F_y$

No

No

Yes

$\dfrac{\ell_b}{r_T} \geqslant \sqrt{\dfrac{102,000 \, C_b}{F_y}}$ No

Sec. 1.5.1.4.5(2)

$\dfrac{\ell_b}{r_T} \geqslant \sqrt{\dfrac{510,000 \, C_b}{F_y}}$

Sec. 1.5.1.4.5(2)

Yes

$F_b'' = \left[\dfrac{2}{3} - \dfrac{F_y (\ell_b/r_T)^2}{1,530,000 \, C_b} \right] F_y \leqslant 0.6 \, F_y$

Sec. 1.5.1.4.5(2)

$F_b'' = \dfrac{170,000 \, C_b}{(\ell_b/r_T)^2} \leqslant 0.6 \, F_y$

Sec. 1.5.1.4.5(2)

F_{bx} = larger value of F_b', F_b''

Flowchart 3.1 *(continued)*

(2) Allowable Bending Stress About Minor Axis (AISCS)

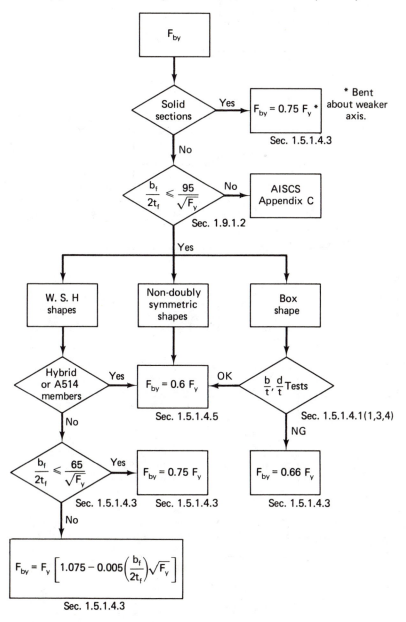

Flowchart 3.2

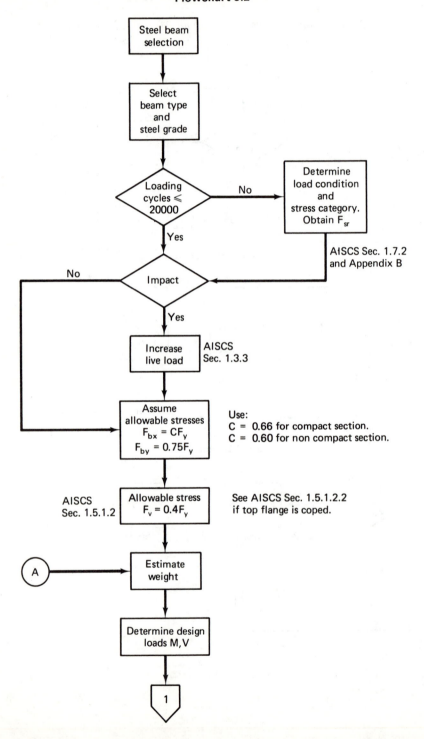

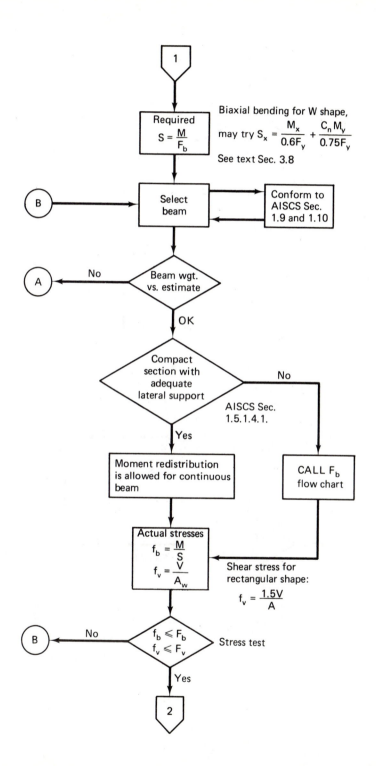

Biaxial bending for W shape,

may try $S_x = \dfrac{M_x}{0.6F_y} + \dfrac{C_n M_y}{0.75F_y}$

See text Sec. 3.8

Required
$S = \dfrac{M}{F_b}$

Select beam

Conform to AISCS Sec. 1.9 and 1.10

B

Beam wgt. vs. estimate — No → A

OK

Compact section with adequate lateral support — No

AISCS Sec. 1.5.1.4.1.

Yes

Moment redistribution is allowed for continuous beam

CALL F_b flow chart

Actual stresses
$f_b = \dfrac{M}{S}$
$f_v = \dfrac{V}{A_w}$

Shear stress for rectangular shape:
$f_v = \dfrac{1.5V}{A}$

$f_b \leqslant F_b$
$f_v \leqslant F_v$ — No → B

Stress test

Yes

2

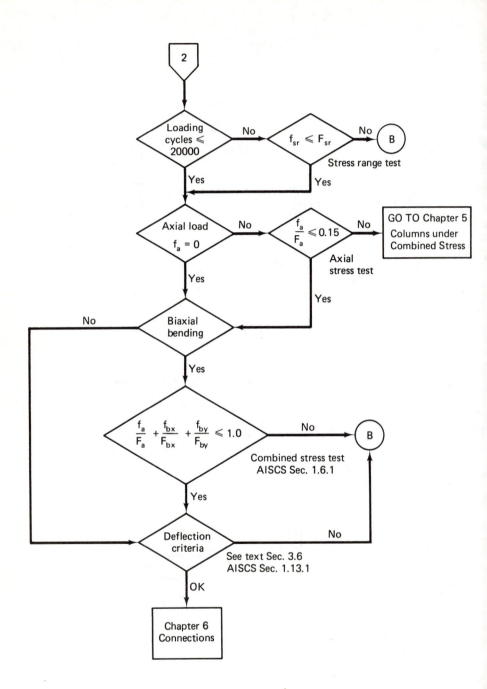

3.12 ILLUSTRATIVE EXAMPLES

The following illustrative problems demonstrate the use of the AISCS and AISCM in the selection and design of steel beams under some of the conditions that have been covered in this chapter. The problems are intended to do more than merely illustrate procedures; they also show how one may change conditions in some cases to achieve greater economy by altering certain details. The problems, in some cases, demonstrate the logic and judgment that may be applied to "zero in" on a solution when the first try turns out to be unsatisfactory. The reader is urged to make up hypothetical design situations on his own and carry out similar solutions.

Example 3.1

A simply supported beam (below) with a span of 20 ft is to carry a static uniform live load of 2.5 kips/ft. The flange is laterally supported by the floor system that it supports. Select the most economical W shape, using A36 steel.

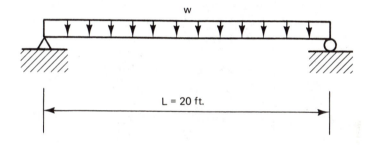

Solution

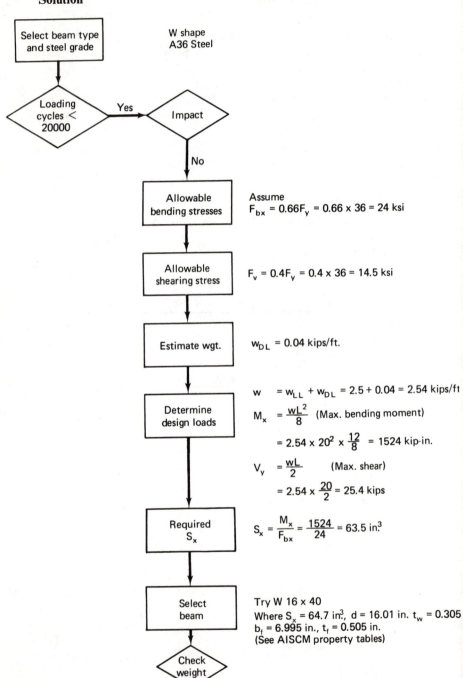

Select beam type and steel grade → W shape A36 Steel

Loading cycles < 20000 — Yes → Impact — No

Allowable bending stresses

Assume
$F_{bx} = 0.66F_y = 0.66 \times 36 = 24$ ksi

Allowable shearing stress

$F_v = 0.4F_y = 0.4 \times 36 = 14.5$ ksi

Estimate wgt.

$w_{DL} = 0.04$ kips/ft.

Determine design loads

$w = w_{LL} + w_{DL} = 2.5 + 0.04 = 2.54$ kips/ft

$M_x = \dfrac{wL^2}{8}$ (Max. bending moment)

$\quad = 2.54 \times 20^2 \times \dfrac{12}{8} = 1524$ kip-in.

$V_y = \dfrac{wL}{2}$ (Max. shear)

$\quad = 2.54 \times \dfrac{20}{2} = 25.4$ kips

Required S_x

$S_x = \dfrac{M_x}{F_{bx}} = \dfrac{1524}{24} = 63.5$ in.^{3}

Select beam

Try W 16 x 40
Where $S_x = 64.7$ in.^{3}, d = 16.01 in. $t_w = 0.305$
$b_f = 6.995$ in., $t_f = 0.505$ in.
(See AISCM property tables)

Check weight — OK

1

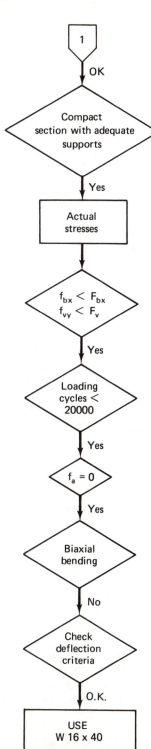

AISCS Sec. 1.5.1.4.1

(1) and (5) are OK

(2) $\dfrac{b_f}{2t_f} = \dfrac{6.995}{2 \times 0.505} = 6.93 < 10.8$ OK

(4) $\dfrac{d}{t_w} = \dfrac{16.01}{0.305} = 52.5 < 106.7$ OK

$f_{bx} = \dfrac{M_x}{S_x} = \dfrac{1524}{64.7} = 23.6$ ksi

$f_{vy} = \dfrac{V_y}{A_w} = \dfrac{25.4}{16.01 \times 0.305} = 5.20$ ksi

AISCS commentary recommendation:

$\dfrac{\ell}{d} < \dfrac{800}{F_y}$ (See text Sec. 3.6)

$\dfrac{\ell}{d} = \dfrac{20 \times 12}{16.01} = 15 < 22.2$ OK

Example 3.2

Same as Example 3.1, but with intermittent lateral supports to provide equivalent full lateral support.

Solution

AISCS, Sec. 1.5.1.4.1(5), provides criteria for the maximum laterally unsupported span segment of compression flange:

$$\frac{76b_f}{\sqrt{F_y}} = 12.7b_f = 12.7 \times 7 = 89 \text{ in.}$$

$$\frac{20,000}{(d/A_f)F_y} = \frac{556}{d/A_f} = \frac{556}{4.53} = 122.7 \text{ in.}$$

Hence the maximum allowable unsupported span segment of the compression flange is 89 in. The top flange is entirely in compression. Provide supports to divide the span in three equal-length segments. Then

$$l_b = \tfrac{240}{3} = 80 \text{ in.} < 89 \qquad \text{OK}$$

Example 3.3

Same as Example 3.1, but without any lateral support.

Solution

Note: A useful guideline that may be used in conjunction with the section modulus (S) requirement is the determination of the maximum value of d/A_f that would not require reduction below $F_{bx} = 0.6F_y$, or, in the case of A36 steel, 22 ksi. Thus for $F_b = 22$ ksi, $l_b = 240$ in., and $C_b = 1$, from AISCS, Sec. 1.5.1.4.5(2), Formula (1.5-7),

$$\frac{d}{A_f} = \frac{12,000C_b}{F_b l_b} = \frac{12,000 \times 1}{22 \times 240} = 2.27 \qquad \text{(max).}$$

Referring to the Elastic Section Modulus Tables of AISCM, and to d/A_f as tabulated in AISCM:

Shape	S_x	d/A_f
W 14 × 53	77.8	2.62
W 16 × 45	72.7	4.06
W 12 × 53	70.6	2.10
W 14 × 48	70.3	2.89

The choices above all provide an adequate section modulus. This tabulation indicates that a rather drastic reduction in allowable bending stress (F_b) would be required for all except the W 12 × 53 shape, for which no reduction at all is necessary. However in other cases greatest economy may involve use of a reduced allowable stress.

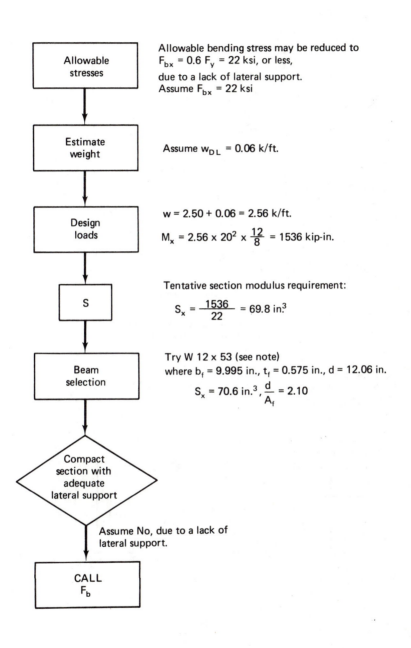

Allowable stresses

Allowable bending stress may be reduced to $F_{bx} = 0.6\ F_y = 22$ ksi, or less, due to a lack of lateral support.
Assume $F_{bx} = 22$ ksi

Estimate weight

Assume $w_{DL} = 0.06$ k/ft.

Design loads

$w = 2.50 + 0.06 = 2.56$ k/ft.

$M_x = 2.56 \times 20^2 \times \dfrac{12}{8} = 1536$ kip-in.

S

Tentative section modulus requirement:

$$S_x = \dfrac{1536}{22} = 69.8 \text{ in.}^3$$

Beam selection

Try W 12 x 53 (see note)
where $b_f = 9.995$ in., $t_f = 0.575$ in., d = 12.06 in.

$$S_x = 70.6 \text{ in.}^3, \dfrac{d}{A_f} = 2.10$$

Compact section with adequate lateral support

Assume No, due to a lack of lateral support.

CALL F_b

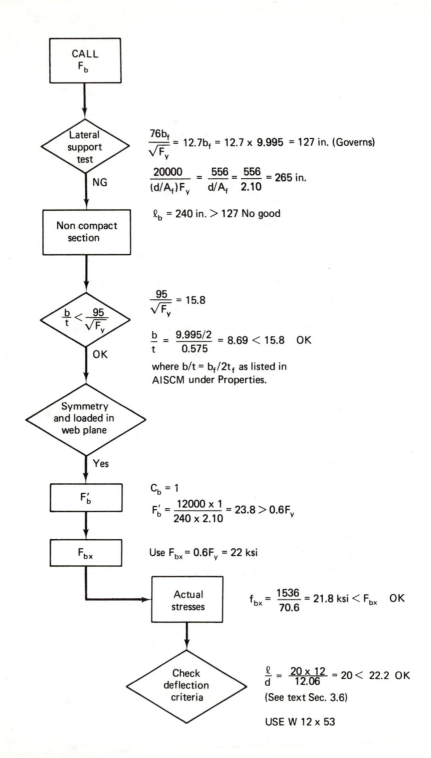

CALL F_b

Lateral support test — NG

$$\frac{76 b_f}{\sqrt{F_y}} = 12.7 b_f = 12.7 \times 9.995 = 127 \text{ in. (Governs)}$$

$$\frac{20000}{(d/A_f)F_y} = \frac{556}{d/A_f} = \frac{556}{2.10} = 265 \text{ in.}$$

$$\ell_b = 240 \text{ in.} > 127 \text{ No good}$$

Non compact section

$$\frac{95}{\sqrt{F_y}} = 15.8$$

$\dfrac{b}{t} < \dfrac{95}{\sqrt{F_y}}$ — OK

$$\frac{b}{t} = \frac{9.995/2}{0.575} = 8.69 < 15.8 \quad \text{OK}$$

where $b/t = b_f/2t_f$ as listed in AISCM under Properties.

Symmetry and loaded in web plane — Yes

F_b'

$C_b = 1$

$$F_b' = \frac{12000 \times 1}{240 \times 2.10} = 23.8 > 0.6 F_y$$

F_{bx}

Use $F_{bx} = 0.6 F_y = 22 \text{ ksi}$

Actual stresses

$$f_{bx} = \frac{1536}{70.6} = 21.8 \text{ ksi} < F_{bx} \quad \text{OK}$$

Check deflection criteria

$$\frac{\ell}{d} = \frac{20 \times 12}{12.06} = 20 < 22.2 \text{ OK}$$

(See text Sec. 3.6)

USE W 12 x 53

Example 3.4

Same beam selection as in Example 3.3, with uniform load replaced by an equivalent concentrated load at midspan that may be repeated 3,000,000 times.

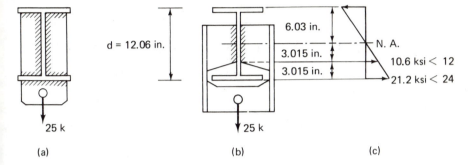

(a) (b) (c)

Solution

A midspan concentrated load of 25 kips causes the same moment as the uniform load of 2.5 kips/ft. The load is to be suspended with initial design of the loading detail in sketch (a).

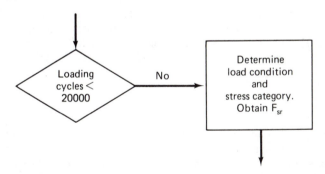

Loading condition is No. 4, as listed in AISCS, Appendix B, Table B1, for repetitions of load exceeding 2,000,000 times. Referring to Table B2, the situation is considered to be one of flexural stress adjacent to a welded transverse stiffener. (Illustrative Ex. 7 of Fig. B1.) The *stress category* is now established as C, and from Table B3, Appendix B, the *allowable stress range* (F_{sr}) is found to be

$$F_{sr} = 12 \text{ ksi}$$

$$M_x = \frac{PL}{4} = \frac{25 \times 20 \times 12}{4} = 1500 \text{ kip-in.} \quad \text{(max. live-load moment)}$$

$$f_{bx} = \frac{1500}{70.6} = 21.2 \text{ ksi} \quad \text{(W 12} \times 53\text{)}$$

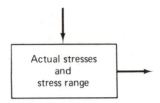

(For design for repeated load, according to AISCS, Appendix B, the stress range $f_{sr} = 21.2 - 0 = 21.2$ ksi in this case. The *stress range* is the algebraic difference between the maximum and minimum stresses caused by live load.)

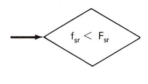

$f_{sr} = 21.2$ ksi $> F_{sr} = 12$ ksi   No good

Thus the design as it stands is unsatisfactory. Rather than select a heavier beam section, the problem can be solved by changing the weld detail as shown in sketch (b) to reduce the live-load stress level adjacent to the transverse weld to something less than F_{sr} (12 ksi), as shown in sketch (c). Then the unwelded as-rolled beam surface at the location for maximum tension stress due to flexure has a stress category of A (Table B2) with an allowable stress range (F_{sr}) of 24 ksi. Category C still holds for the web at the stiffener and $F_{sr} = 12$ ksi. The beam design is OK. The design of the welded connection is deferred to Chapter 6.

Example 3.5

A simply supported beam with a span of 20 ft carries a uniform dead load of 1 kip/ft, including its own weight. A concentrated oblique load at midspan acts through the centroid, as shown. Determine economical beam selection for A36 steel if there are no intermediate lateral supports.

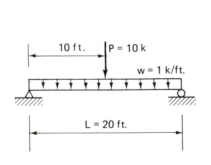

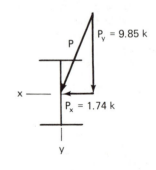

Solution

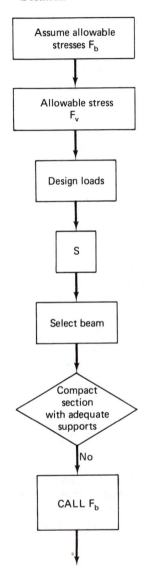

$F_{bx} = 0.6F_y = 22$ ksi
$F_{by} = 0.75F_y = 27$ ksi

$F_v = 0.4F_y = 14.5$ ksi

$$M_x = \frac{wL^2}{8} + \frac{P_y L}{4}$$

$$= 1 \times \frac{20^2}{8} + 9.85 \times \frac{20}{4} = 99.25 \text{ k-ft. or } 1191 \text{ k-in.}$$

$$M_y = \frac{P_x L}{4} = 1.74 \times \frac{20}{4} = 8.7 \text{ k-ft. or } 104 \text{ k-in.}$$

$$V_x = 0.87 \text{ k,} \qquad V_y = 14.9 \text{ k}$$

$$S_x = \frac{M_x}{0.6F_y} + \frac{C_n M_y}{0.75F_y} = \frac{1191}{0.6 \times 36} + \frac{5 \times 104}{0.75 \times 36} = 74.4 \text{ in.}^3$$

(See text Sec. 3.8)

Try W 12 x 53
where $S_x = 70.6$ in.^{3}, $S_y = 19.2$ in.^{3}

$F_{bx} = 0.60F_y = 22$ ksi (as in Example 3.3)
$F_{by} = 0.75F_y = 27$ ksi,

$$\frac{65}{\sqrt{F_y}} = 10.8$$

$$\frac{b_f}{2t_f} = \frac{9.995}{2 \times 0.575} = 8.69 < 10.8 \quad \text{OK}$$

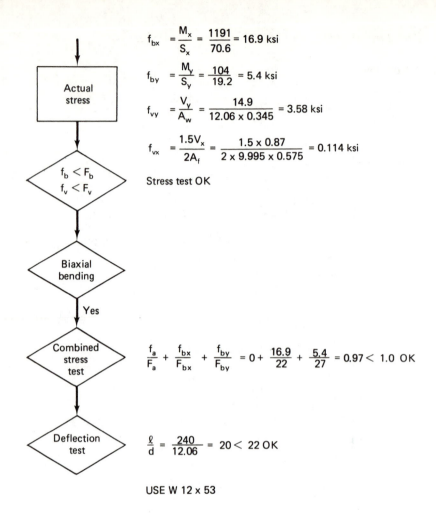

$$f_{bx} = \frac{M_x}{S_x} = \frac{1191}{70.6} = 16.9 \text{ ksi}$$

$$f_{by} = \frac{M_y}{S_y} = \frac{104}{19.2} = 5.4 \text{ ksi}$$

$$f_{vy} = \frac{V_y}{A_w} = \frac{14.9}{12.06 \times 0.345} = 3.58 \text{ ksi}$$

$$f_{vx} = \frac{1.5V_x}{2A_f} = \frac{1.5 \times 0.87}{2 \times 9.995 \times 0.575} = 0.114 \text{ ksi}$$

Actual stress

$f_b < F_b$ $f_v < F_v$

Stress test OK

Biaxial bending

Yes

Combined stress test

$$\frac{f_a}{F_a} + \frac{f_{bx}}{F_{bx}} + \frac{f_{by}}{F_{by}} = 0 + \frac{16.9}{22} + \frac{5.4}{27} = 0.97 < 1.0 \text{ OK}$$

Deflection test

$$\frac{\ell}{d} = \frac{240}{12.06} = 20 < 22 \text{ OK}$$

USE W 12 x 53

Example 3.6

A simple span crane runway beam supports a moving load, transmitted by two wheels, as shown, of 80 kips, including impact. Using an unsymmetrical section of the type shown in Fig. 3.12(b), determine required plate sizes. Assume that the 16-kip lateral load is applied at the level of the top flange. Use ASTM A36 steel.

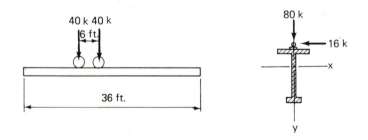

Solution

This problem will illustrate successive "cut-and-try" steps in approaching a final economical section. As discussed in Section 3.8, it will be assumed that all the lateral load is carried by the top flange to compensate for the fact that torsional stresses will not be considered.

As a preliminary guide to determining a suitable size for the top (compression) flange, we make an initial "guess" that the total stress due to the sum of bending about the x and y axes will be 22 ksi and that in the compression flange about 75% is due to the lateral load; that is, f_{by} in compression is about 16.5 ksi.

As shown in many texts in structural analysis, the maximum moment due to a moving two-wheel load system is under the wheel nearest the center of the span when it and the center of gravity of the moving load system are equidistant from the center of the span.

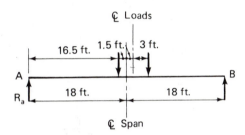

Calculate the maximum moments due to vertical and lateral live loads:

$$R_{ax} = 16\frac{16.5}{36} = 7.33 \text{ kips}$$

$$R_{ay} = 80\frac{16.5}{36} = 36.65 \text{ kips}$$

$$M_x = 16.5R_{ay} = 604.73 \text{ kip-ft} = 7260 \text{ kip-in.}$$

$$M_y = 16.5R_{ax} = 121 \text{ kip-ft} = 1452 \text{ kip-in.}$$

The section modulus for the top flange alone is

$$S_{yf} = \frac{b_f^2 t_f}{6}$$

$$f_{by} = \frac{M_y}{S_{yf}} = \frac{1452 \times 6}{b_f^2 t_f}, \qquad b_f = \sqrt{\frac{1452 \times 6}{f_{by}t_f}}$$

Try $t_f = 1.0$ in. and $f_{by} = 16.5$ ksi (assumed); then

$$b_f = \sqrt{\frac{1452 \times 6}{16.5 \times 1}} = 22.97 \text{ in.}$$

A. *First try:*

Top flange:	*PL* 1 × 22
Bottom flange:	*PL* 1 × 8
Web:	*PL* ½ × 34

Remarks:

1. A trial overall depth of $\frac{1}{12}$ of the span, or $d = 36$ in., will be used.
2. $h/t = 68$, exceeding 63.3, as listed in "Allowable Stress in Shear" in Section 3.4, but shear stress will probably be well under 14.5 ksi.
3. If too narrow, bottom flange may buckle during erection, or vibrate during crane operation.

1. *Section properties:*

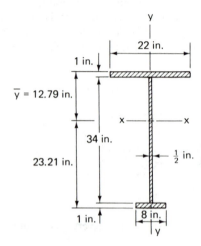

Locate the neutral axis; take the area moments about the center of the top flange:

Top flange:	22.0×0	$=$	0
Bottom flange:	8.0×35.0	$=$	280.0
Web:	17.0×17.5	$=$	297.5
Total	47.0		577.5

$$\bar{y} = \frac{577.5}{47.0} + \frac{1}{2} = 12.79 \text{ in.} \qquad \text{(see the figure)}$$

Determine the moment of inertia I_x and the section moduli (neglect I_o of the flanges):

Top flange:	22.0×12.29^2	$= 3323$
Bottom flange:	8.0×22.71^2	$= 4126$
Web:	17.0×5.21^2	$= 461$
	$34.0^3 \times 0.5/12$	$= 1638$
Total		$I_x = 9548 \text{ in.}^4$

$$S_{xc} = \frac{9548}{12.79} = 746.5 \text{ in.}^3 \qquad \text{for compression flange}$$

$$S_{xt} = \frac{9548}{23.21} = 411.4 \text{ in.}^3 \qquad \text{for tension flange}$$

$$S_{yf} = \frac{22^2 \times 1.0}{6} = 80.7 \text{ in.}^3 \qquad \text{for compression flange alone, about } y \text{ axis}$$

2. *Determine the allowable stress* (F_{bx}) *for the laterally unsupported top flange (see the F_b flowchart):*

$$F'_{bxc} = \frac{12,000C_b}{l(d/A_f)} = \frac{12,000 \times 1}{(36 \times 12)[36/(22 \times 1)]} = 17 \text{ ksi} < 22 \qquad \text{OK}$$

$$\text{top flange } I_y = \frac{22^3 \times 1}{12} = 887.3 \text{ in.}^4$$

$$\tfrac{1}{3} \text{ comp. area of web} = \frac{11.79 \times 0.5}{3} = 1.97 \text{ in.}^2$$

$$r_T = \sqrt{\frac{887.3}{23.97}} = 6.08 \text{ in.} \qquad \text{[AISCS, Sec. 1.5.1.4.5(2)]}$$

$$\frac{l}{r_T} = \frac{36 \times 12}{6.08} = 71.1$$

[Use AISCS, Formula (1.5-6a) with $C_b = 1$, since $53 < l/r_T < 119$.]

$$F''_{bxc} = 24.0 - \frac{71.1^2}{1181} = 19.72 \text{ ksi} < 22 \qquad \text{OK}$$

Use larger value of F'_{bxc} and F''_{bxc}; then $F_{bxc} = 19.72$ ksi.

3. Determine the stresses due to bending of the top flange (live load):

$$f_{bxc} = \frac{M_x}{S_{xc}} = \frac{7260}{746.5} = 9.7 \text{ ksi}$$

$$f_{byc} = \frac{M_y}{S_{yf}} = \frac{1452}{80.7} = 18.0 \text{ ksi}$$

At this point, since the total combined compressive stress in the top flange is 27.7 ksi, it is obvious that a new trial section will need to be chosen. However, as a guideline to the modification that may be needed, the stress due to tension will be calculated.

$$f_{bx} = \frac{M_x}{S_{xt}} = \frac{7260}{411.4} = 17.65 \text{ ksi}$$

B. *Choose the second trial section:* Note that the neutral axis may be moved up and that the compression area and/or the depth of the section needs to be increased. Also, F_{bxc} may as well be increased by widening the top flange. The stress due to dead load, previously omitted as a minor factor in preliminary calculations, will now be introduced. Omitting calculations that are a repetition in form of those made for the first trial, the following are found for the second trial section:

Top flange:	PL 1 × 26
Bottom flange:	PL 1 × 7
Web:	PL $\tfrac{1}{2}$ × 36

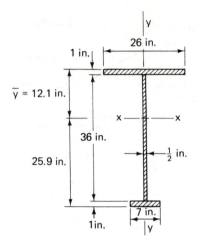

$$I_x = 10{,}995 \text{ in.}^4$$

$$S_{xc} = 908.7 \text{ in.}^3$$

$$S_{xt} = 424.5 \text{ in.}^3$$

$$S_{yf} = 112.7 \text{ in.}^3 \qquad \text{(top flange alone)}$$

$$r_T = 7.25 \text{ in.}$$

$$\frac{l}{r_T} = 59.6$$

$$F_{bxc} = 21.0 \text{ ksi}$$

Dead load moment (unit weight $= 51.0 \times 3.4 = 173 \text{ lb/ft}$):

$$M_{DL} = \frac{0.173 \times 36^2 \times 12}{8} = 337 \text{ kip-in.}$$

(Note that while the maximum live load moment occurs 1.5 ft from the center of the span, at which point the dead load moment is slightly less than 337 kip-in., it will be a simple, conservative, and closely approximate procedure to add the two maximum values.)

Combining the maximum moments due to live and dead load,

$$M_x = 7260 + 337 = 7597 \text{ kip-in.}$$

The stresses due to bending are

$$f_{bxc} = \frac{M_x}{S_{xc}} = \frac{7597}{908.7} = 8.36 \text{ ksi}$$

$$f_{byc} = \frac{M_y}{S_{yf}} = \frac{1452}{112.7} = 12.88 \text{ ksi}$$

The combined compressive stress is now checked by the AISCS interaction formula (1.6-2), with $f_a = 0$ (AISCS, Sec. 1.6.1):

$$\frac{f_a}{F_a} + \frac{f_{bx}}{F_{bx}} + \frac{f_{by}}{F_{by}} = 0 + \frac{8.36}{21.0} + \frac{12.88}{22.0} = 0.98 < 1.0 \qquad \text{OK}$$

(Note that by AISCS, Sec. 1.5.1.4.3, assuming compliance with Sec. 1.5.1.4.1(1) and (2), an allowable stress of 27 ksi could not in any case be used because the section is not doubly symmetric.)

The stress in tension flange due to bending is

$$f_{bxt} = \frac{M_x}{S_{xt}} = \frac{7597}{424.5} = 17.9 \text{ ksi} < 22 \text{ ksi} \qquad \text{OK}$$

Check the shear stress. For this purpose, the moving wheel loads are placed at the extreme left end to produce the maximum shear, assumed equal to the maximum reaction:

$$V_y = \frac{80 \times 33}{36} + 0.173 \times 18 = 76.4 \text{ kips}$$

$$f_v = \frac{V_y}{A_w} = \frac{76.4}{18} = 4.24 \text{ ksi} < 14.5 \qquad \text{OK}$$

(The shear stress is so much below the allowable there is no point in checking by the more accurate VQ/It formula.)

Check the width/thickness ratios. For the flange:

$$\frac{b}{t} = \frac{13.0}{1.0} = 13$$

which is less than the 15.8 allowed for outstanding unstiffened elements (AISCS, Sec. 1.9.1.2). For the web:

$$\frac{h}{t} = \frac{36}{0.5} = 72$$

This exceeds the maximum value of 63.3, as tabulated in Section 3.4. However, for cases where the maximum shear stress is less than F_v for the case of no intermediate stiffeners (AISCS, Sec. 1.10.5.2, with $a/h = \infty$), it is permissible for h/t to exceed the values tabulated in Section 3.4. For a more complete discussion of this topic, reference should be made to Chapter 7, where treatment of the problem includes the use of intermediate stiffeners. At this time, for design of girders without intermediate stiffeners, use may be made of the extreme right column captioned "over 3" in Table 10-36 of AISCS, Appendix A. This column lists the maximum permissible shear stress for girders without intermediate stiffeners. In the present example, interpolating between the value of 13.0 for $h/t = 70$ and 11.4 for $h/t = 80$, the maximum allowable stress for $h/t = 72$ would be $F_v = 12.7$ ksi. Since this exceeds the maximum average shear stress of 4.24 ksi, the design is satisfactory.

PROBLEMS

3.1. Assuming that full lateral support is provided, determine the maximum moment and shear in each of the following cases and make the most economical beam selection. Neglect beam dead weight in initial selection, then check stress due to both live and dead weight. The span is 24 ft in each case. The beams are simply supported at each end. The steel is ASTM A36 with $F_y =$ 36 ksi.

    a. Uniform load of 8 kips/ft.

    b. Concentrated center load of 165 kips.

    c. Three loads at $\frac{1}{4}$ points of 75 kips each.

3.2. A simply supported beam with a span of 34 ft carries a static uniform load of 3 kips/ft. The compression flange is laterally supported by the floor slab. Select the most economical W shape for a steel with $F_y = 46$ ksi. (Refer to Example 3.1.)

3.3. Same as Problem 3.2, but with intermittent lateral supports, as may be required, to provide the equivalent of full lateral support. (See Example 3.2.)

3.4. Same as Problem 3.2, but without any lateral support. (See Example 3.3.)

3.5. Design a simple beam for a span of 24 ft, using A36 steel, under a uniform load of 4 kips/ft and a concentrated load of 90 kips 2 ft from one end. Intermittent lateral supports are provided at 4-ft intervals.

3.6. Design a simple bearing block support for the heavily loaded end of the beam selected in Problem 3.5. Refer to Section 3.9.

3.7. For the beam selected in Problem 3.5, compare (a) the maximum shear stress at the end of the beam nearest the concentrated load, using Eq. (3.8), (b) the average shear stress at the same location by AISCS, and (c) the average shear stress based on the clear depth web area, that is, the web area between the inner faces of the flanges.

3.8. Same as Problem 3.1, but with only a single lateral support at midspan.

3.9. Same as Problem 3.1, but with no lateral support between the ends of the beam.

3.10. Same as Problem 3.1(b), but with the 165-kip load repeated 1,200,000 times. Beam is fully supported laterally and it is assumed that the concentration of load at the center will require the use of a vertical stiffener similar to that sketched in Case 7 of AISCS. Use Example 3.4 as a guide. If the stress range permitted by the repeated load requirements controls the design, give special attention to a modification of stiffener details to improve the stress category as may be required.

3.11. A standard rectangular structural tube is used to span a highway with an effective simple span of 42 ft. In addition to its own dead weight, it supports a highway road direction sign having a total weight of 2000 lb, which is attached to the tube at its third points. The sign, measuring 4 ft in height by 18 ft in length, is centrally located and transmits to the tube a horizontal wind force that may be as great as 40 psf. This force is also introduced at the third points. Select a tube, using A36 steel. Do not neglect the wind force on the tube alone at each end of the span.

3.12. By use of Eq. (3.9) and the values of section moduli (S) and plastic moduli (Z) as provided in the AISCM, determine the shape factor for the following sections:

W 36 × 300

W 18 ×   40

S 24 ×   80

W 14 × 605

3.13. Using A36 steel, select the required size for the simply supported beam with overhang loaded as shown in the figure. Note that lateral support is provided only at the location of the 40-kip load and at the reaction supports.

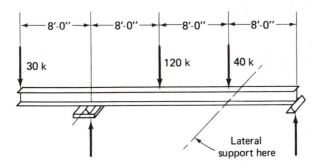

3.14. Select a W shape, using A36 steel, for a span of 36 ft, to support a vertical load of 500 lb/ft in addition to its own dead weight. The beam, being exposed must also resist a horizontal wind force of 30 psf and is laterally unsupported.

3.15. A simple beam with a span of 24 ft carries a uniform vertical load of 1 kip/ft, including dead weight of beam, and a concentrated load at the center of 80 kips. Using A242 steel select:
   a. A W shape for the vertical load alone.
   b. A W shape for the vertical load plus a concentrated horizontal force of 6 kips, at the center of the span, and applied at the neutral axis of the beam.
   c. Same as part (b), but with the horizontal force applied at the level of the top flange. (Assume that the top flange alone acts as a rectangular beam to resist the lateral force and keep the combined compressive stress within the allowable value of F_b.)

3.16. Same as Problem 3.15(c), but using a built-up welded member made of three plates similar to the section used in Example 3.6.

3.17. Similar to Example 3.6, except change span to 42 ft, total wheel load (without impact) 36 kips vertical and 12 kips horizontal. Apply the AISCS impact allowance for crane runway support girders to both the vertical and horizontal loads. The two wheel locations are 8 ft apart. Use A36 steel. Study Example 3.6 before attempting this problem, and try to use it as a guide in making a better initial selection of flange plate sizes.

4

COLUMNS UNDER AXIAL LOAD

4.1 INTRODUCTION

Originally, the term "column" referred to a vertically upright compression member, such as are found in Egyptian, Greek, or Roman temples, built of hand-hewn segments of rock or marble. In today's usage, a column is not necessarily upright, and any compression member, horizontal, vertical, or inclined, is termed a *column* if the compression that it transmits is the primary factor determining its structural behavior. If bending is also a major factor, the term *beam-column* may be used, and such members will be considered in Chapter 5.

For engineering design purposes, the *axially loaded* column is defined as one that transmits a compressive force whose resultant at each end is approximately coincident with the longitudinal centroidal axis of the member. Although there are no design loads that produce bending moment, there may be moments due to initial imperfections, accidental curvature, or unintentional end eccentricity. Such accidental bending moments reduce the strength of the member, but are assumed to be taken care of in the design formula by an appropriate factor of safety.

The use of the column in structures has at times jumped ahead of design knowhow. The tragic failure of the first attempt to build the Quebec Bridge, in 1907, has been attributed to faulty column design in which proportions and sizes were extrapolated beyond the range of previous experience. Unfortunately, the failure of compression members in structures is still too prevalent. An understanding of column behavior is most important to the structural engineer, as it provides an aid to the intelligent use of design specifications.

Column failure involves the phenomenon of *buckling*, during which a member experiences deflections of a totally different character than those associated with the initial loading. Thus, when an axially loaded column is first loaded, it simply shortens or compresses in the direction of the load. If and when a buckling load is reached, the shortening deformation stops, and a sudden lateral and/or twisting deformation occurs in a direction normal to the column axis, thus limiting the axial load capacity.

The strength of a tension member is independent of its length, whereas for the column both the strength and the mode of failure are markedly dependent on length. A very short and compact column built of any one of the common metals will develop about the same strength in compression as it will in tension. But if the column is long, it will fail at a load that is proportional to the bending rigidity of the member, *EI*, and independent of the strength of the material. Thus a very slender column of steel with a yield stress of 100 ksi has no more column strength than one with a yield stress of 36 ksi. In each case the strength is determined by the Euler column formula that was developed more than 220 years ago.

4.2 BASIC COLUMN STRENGTH

The buckling strength of a column decreases with increasing length. Beyond a certain length the buckling stress falls below the proportional limit of the material and buckling for any longer column is elastic. For such a slender column the buckling load is given by the Euler formula,

$$P_e = \frac{\pi^2 EI}{l^2} \tag{4.1}$$

The yield stress, F_y, does not appear in Eq. (4.1). It plays no part in determining the strength of a very long column. Thus a slender aluminum alloy column, according to Eq. (4.1), will buckle at about one-third of the load of its steel counterpart—not because of any weakness in the material, but simply because the elastic modulus E of aluminum alloy is about one-third that of steel. The aluminum column would also weigh about one-third that of steel, and by reshaping the design of the cross section the strength can be increased, up to a limit, with *no increase in the weight of the member*. The limit of such increase occurs when the material gets so spread out, and correspondingly thin, that *local buckling* occurs prior to general buckling of the entire member.

The Euler load, P_e, is a load that will just hold the column in the deflected shape shown in Fig. 4.1. At any point along the column the external applied moment Py is equal to the internal resisting moment, $EI\phi$, where ϕ is the column curvature at the corresponding point. (Refer to Section 3.2)

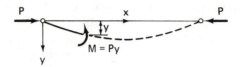

Fig. 4.1 Buckled shape of a hinged-end column.

If both sides of Eq. (4.1) are divided by A and the relationship $I = Ar^2$ is introduced, r being the radius of gyration of the cross section, the buckling load is expressed in terms of the buckling stress, F_e:

$$F_e = \frac{P_e}{A} = \frac{\pi^2 EI}{Al^2} = \frac{\pi^2 E}{(l/r)^2} \tag{4.2}$$

Equation (4.2) can be modified so as to apply to other end conditions, such as free or fixed, by use of the effective length factor K. For purely flexural buckling, Kl is the length between inflection points and is known as the effective length. Thus Eq. (4.2) becomes

$$F_e = \frac{\pi^2 E}{(Kl/r)^2} \tag{4.3}$$

For example, if a column is fixed against rotation and translation (lateral movement) at each end, it will buckle with points of inflection at the quarter points, as shown in Fig. 4.2(a), and the effective length factor is 0.5. Consequently, according to Eq. (4.3), a very slender column with fixed ends that

Fig. 4.2 Example illustrating the effective length concept.

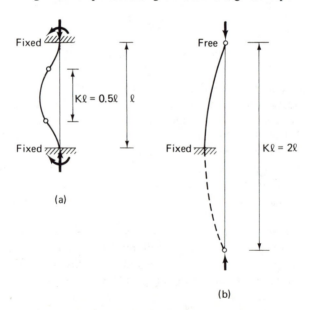

(a)

(b)

buckles elastically will be four times stronger than the same column with hinged ends. But if one end is fixed and the other free both with respect to rotation and translation, as shown in Fig. 4.2(b), there is an imaginary point of inflection at a distance *l* below the column base, and the effective length factor is 2.0. Such a column has only one-fourth the *elastic* strength of the same member with hinged ends. Table 4.1 shows these and other cases and lists recommended modified values for design usage.

Table 4.1

Effective Length Factors K for Centrally Loaded Columns with Various Idealized End Conditions

	(a)	(b)	(c)	(d)	(e)	(f)
Buckled shape of column is shown by dashed line.						
Theoretical K value	0.5	0.7	1.0	1.0	2.0	2.0
Recommended K value when ideal conditions are approximated	0.65	0.80	1.2	1.0	2.10	2.0

End condition code	
	Rotation fixed Translation fixed
	Rotation free Translation fixed
	Rotation fixed Translation free
	Rotation free Translation free

l/r is termed the *slenderness ratio* and is almost universally used as a parameter in terms of which the column-strength curve may be drawn graphically or expressed analytically by a column-strength formula. Figure 4.3 shows typical column-strength curves for steel. The strengths of the very short and very long column are expressed by F_y and F_e, respectively. In the intermediate range, the transition from F_y to F_e depends on a complex mix of factors—initial curvature, accidental end eccentricity, and residual stress—and is usually expressed empirically by means of parabolic, straight-line, or more complex expressions.

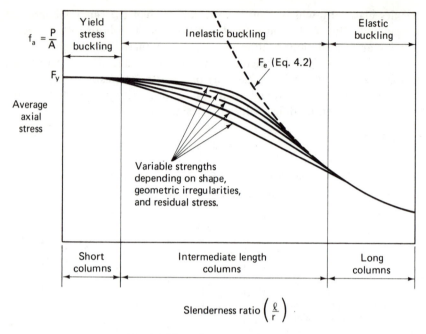

Fig. 4.3 The column strength curve.

In the case of structural steel, the presence of residual stress in rolled shapes has been shown to be an important factor influencing the shape of the transition curve between very short and very long columns.† Residual stresses are locked in a member as a result of uneven cooling after rolling, welding, oxygen cutting, or by cold-straightening operations.

On the basis of column tests, as well as measurements of residual stresses in rolled shapes, the Column Research Council‡ (CRC) proposed in 1960 an empirical transition for the short and intermediate range, from F_y for $Kl/r = 0$, to the Euler curve (F_e) at $F_y/2$ as given by Eq. (4.4). For $Kl/r < C_c$,

$$F_c = \left[1 - \frac{1}{2C_c^2}\left(\frac{Kl}{r}\right)^2\right]F_y \qquad (4.4)$$

where

$$F_c = \text{column strength, ksi}$$

$$C_c = \sqrt{2\pi^2 E/F_y}$$

For $Kl/r \geq C_c$, the Euler formula, Eq. (4.3), should be applied.

†For a more complete treatment of inelastic buckling and the residual stress effect, refer to *The Structural Stability Research Council Guide to Stability Design Criteria for Metal Structures* (Ref. 1.6).

‡Now renamed *Structural Stability Research Council* (SSRC).

In the selection of columns to support design loads, the allowable stress, F_a, is determined by dividing Eq. (4.4) or (4.3) by an appropriate safety factor. This will be discussed in Section 4.7 and will be followed by illustrative design examples.

4.3 EFFECTIVE LENGTH OF COLUMNS

The basic concept of effective length has been explained in the previous section, and certain special cases were shown in Table 4.1. A more general evaluation of effective length factors for columns in continuous frames is available through the use of the alignment charts shown in Fig. 4.4, as presented in the Structural Stability Research Council (SSRC) *Guide*. These charts are functions of the I/L values of adjacent girders (beams), which are assumed to be rigidly attached to the columns. A conservative assumption is made that all columns in the portion of the framework under consideration reach their individual buckling loads simultaneously. The charts are based upon a slope-deflection analysis that includes the effect of column load. In Fig. 4.4(a) and (b) the subscripts A and B refer to the joints at the two ends of the column that is under consideration. G is defined as

$$G = \frac{\sum I_c/L_c}{\sum I_g/L_g} \tag{4.5}$$

In Eq. (4.5) the $\sum$ indicates a summation of all members rigidly connected to that joint (*A or B*) and lying in the plane in which buckling of the column is being considered. I_c is the moment of inertia and L_c is the unsupported length of a column section. I_g is the moment of inertia and L_g is the length of a girder (beam) or other restraining member. I_c and I_g are taken about the axis perpendicular to the plane of buckling.

In connection with the use of these charts, the following recommendations are made by the SSRC *Guide*. For a column base connected to a footing or foundation by a frictionless hinge, G is theoretically infinite, but should be taken as 10 in design practice. If the column base is rigidly attached to a properly designed footing, G approaches a theoretical value of zero, but should be taken as 1.0.

For greater accuracy, the girder stiffness I_g/L_g in Eq. (4.5) should be multiplied by a factor when certain conditions at the far end are known to exist. For the cases with sidesway prevented [Fig. 4.4(a)], the appropriate multiplying factors are 1.5 for far end of girder (beam) hinged, and 2.0 for far end of girder fixed. For the case with sidesway not prevented [Fig. 4.4(b)], the multiplying factors are 0.5 for far end of girder hinged, and 0.67 for far end of girder fixed.

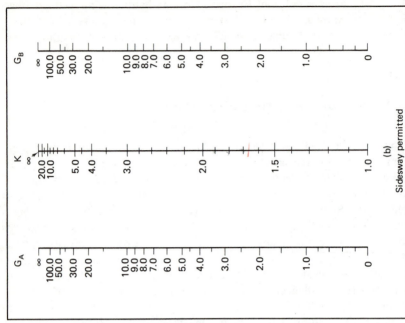

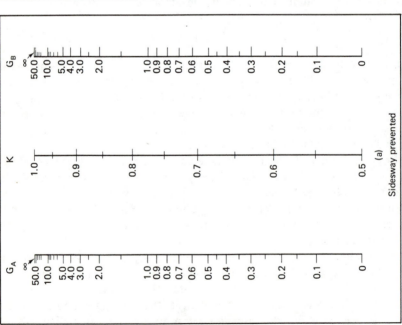

Fig. 4.4 Charts for effective length of columns in continuous frames. (From the Structural Stability Research Council *Guide*, therein by courtesy of Jackson and Moreland Division of United Engineers and Constructors,

Having determined G_A and G_B for a column, K is obtained by constructing a straight line between the appropriate points on the scales for G_A and G_B. For example, in Fig. 4.4(a) if G_A is 0.5 and G_B is 1.0, then K is found to be 0.73.

4.4 TYPES OF STEEL COLUMNS

Various cross-sectional column shapes are shown in Fig. 4.5. The column cross section to be used will be conditioned by the magnitude of the load and by the type of end framing or connections that are most convenient for the particular structural application. Thus a pipe column with top and bottom

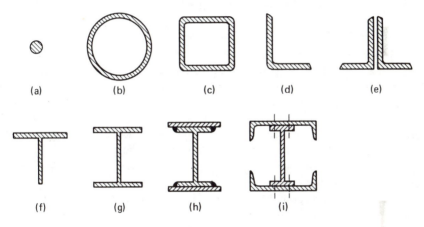

(a)   (b)   (c)   (d)   (e)

(f)   (g)   (h)   (i)

Fig. 4.5 Types of steel columns.

bearing plates and based on a concrete footing is excellent as an isolated unit supporting a beam, but would not be as well suited to a truss with gusset plate connections. In general, within the limits of available clearance and with an eye to thickness limitations, the designer uses a section with the largest possible radius of gyration, thus reducing the slenderness ratio and increasing the allowable stress. In building design, of course, with rather small lengths and the need to maximize the available occupancy space, heavy compact sections are used for heavy loads. Similarly, in exposed areas it may be desirable to use a small section to minimize wind loads.

a. Round, Solid Bars

The radius of gyration (r) for a solid round cross section is equal to

$$r = \frac{d}{4} = \frac{A}{\pi d} \qquad (4.6)$$

where

$$d = \text{diameter of the solid round bar}$$

$$A = \text{cross-sectional area of the bar}$$

Solid round bars of high-strength steel have found particular use as main column elements of tall TV or radio towers, which have been built to heights of more than 2000 ft. Aside from its increased strength, high-strength steel reduces dead weight, which is particularly beneficial when seismic forces are being considered. Similarly, the use of a round section of relatively small diameter minimizes wind force and reduces added weight due to ice formation. Obviously, lateral force is a problem in the design of the individual column element acting as a beam, as in an exposed tower structure. The weight and area of round and square solid bars are tabulated in the AISCM.

b. Steel Pipes

The steel pipe as shown in Fig. 4.5(b) is more efficient than the solid round bar, since the radius of gyration may be increased almost independently of the cross-sectional area, thus reducing l/r and increasing the allowable stress, F_a. The possiblity of local buckling must be considered if the wall thickness in comparison to the pipe diameter becomes overly small. The full allowable column stress may be used as long as the D/t ratio is less than $3300/F_y$, (AISCS, Sec. 1.9.2.3). (D is the mean diameter of the pipe and t is the wall thickness.) These limits are tabulated in Table 6, Appendix A, of AISCS for commonly used yield points.

Material cost per pound for tubular shapes almost always exceeds that of standard rolled shapes, and the end connections in a framed structure will require special consideration. Advantages include those listed for solid round bars. If the ends are hermetically sealed to prevent access of air, the pipe interiors need not be treated to prevent corrosion. Reference should be made to the AISCM for dimensions and properties of available standard sizes of steel pipe, ranging from the smallest up to 12 in. in diameter. Such pipe are available with a yield stress of 36 ksi, and all have D/t ratios well within the limit of $3300/F_y$.

c. Box Sections and Structural Tubes

The box section shown in Fig. 4.5(c) is one of the standard types available, square up to 10×10 in. or rectangular up to 12×8 in., with properties of cross section as listed in the AISCM. Larger sizes may be built up by welding various combinations of plates, angles, or channels. As a compression member, the square tube combines the effectiveness of the hollow steel pipe with the advantage of simpler end connection details in usual building-frame applications.

d. Angle Struts

Single-angle struts, as shown in Fig. 4.5(d), are satisfactory as secondary members for light loads. If they can be used in a situation where the load can be brought uniformly into both angle legs at each end or where the ends can be prevented from rotation or twist by a rigid end connection attachment to a heavy member or footing, they may be designed by usual procedures for centrally loaded columns, provided width/thickness ratio limitations are met.

Double-angle struts, as shown in Fig. 4.5(e), are often used in single plane trusses. Frequent "stitching" must be provided by means of bolts or rivets to ensure that the two angles act as a single unit (see AISCS, Sec. 1.18.2.4). End connections should be designed so as to result in uniform distribution of load. The cross-sectional properties of selected double-angle sections are tabulated in the AISCM for members made up either of two unequal leg angles or two equal leg angles.

The AISCS, Sec. 1.9.1.2, provides that single-angle struts as well as double-angle struts with separators subject to axial compression shall be considered as fully effective when the ratio of width to thickness is not greater than $76/\sqrt{F_y}$ (for struts comprising double angles in contact the ratio shall be not greater than $95/\sqrt{F_y}$); otherwise, the allowable compression stress shall be modified by the appropriate reduction factor as provided in the AISCS, Secs. C2 and C5. The foregoing width/thickness limits are tabulated for various yield points in Appendix A of the AISCS, Table 6.

e. Structural Tees

Structural tees (WT shapes) as shown in Fig. 4.5(f), are often used as chord sections in light welded trusses, with double-angle struts welded to the tee web. Structural tees are made by splitting W shapes, or standard beams, longitudinally; hence they have stems (webs) considerably thinner than the flange. The restraint of the flange with respect to buckling of the stem permits the use of greater width/thickness ratios for the stem of the tee in comparison with the angle struts, as also provided by the AISCS, Sec. 1.9.1.2, permitting a width/thickness ratio in the stem up to $127/\sqrt{F_y}$, as tabulated in Appendix A of the AISCS, Table 6. For ratios greater than this limit, the allowable compression stress is modified in accordance with Secs. C2 and C5 of the AISCS.

f. Wide-Flange Shapes

Wide-flange W, M, or HP shapes are doubly symmetric, as shown in Fig. 4.5(g), and are rolled in a wide range of weights and sizes. These are therefore suited to a correspondingly great range of column loads and lengths and find frequent use in building construction. The W 14 series provides a wide range of coverage of shapes (some with exceptionally wide flanges to balance r about the two axes), which are especially suited to column requirements for tall multistory building frames.

In areas remote from mills that produce heavy W shapes, equivalent sections are produced by means of continuous longitudinal welds joining three plate segments and designated WW. Combination shapes may also be made up by means of a cover-plated W member, as shown in Fig. 4.5(h), or an S (standard beam) shape together with two channel (C) shapes. The latter shape might be desirable for a very long member with not so great loads, for which the cross section should be spread as much as possible to reduce the l/r ratio. Although Figs. 4.5(h) and (i) are shown as welded and bolted (or riveted), respectively, either fabrication method could be used for either shape.

Figure 4.6 shows a heavy W 14 column shape shortly after erection in one of the lower floors of an office building. The heavy bolted column splice has been completed at the lower end, and the splice plates at the upper end are ready to receive the next two-story column segment. Note the beam-connecting plates that are welded to the columns and field bolted to the beams.

g. Columns with Lacing, Battens, or Perforated Cover Plates

In a situation where a very long column is required, it may be necessary to spread the cross section to a degree that makes a laced column economical, as shown for example in Fig. 4.7(a). Before the advent of rolled W shapes, such laced members were also used for more usual lengths found in bridges and buildings, where they may still be seen in older structures. Laced columns are used today in derrick booms and TV or radio towers. In such applications the four-angle member shown in Fig. 4.7(a) together with the lacing bars may be replaced by solid round bars with welded end connections, which help reduce wind and icing loads on exposed members.

Lacing bars carry no column load, but they do perform the following functions:

1. They hold the component parts of the laced column in position so as to maintain the shape of the overall column cross section. For the same purpose, cross bracing in a plane normal to the column axis must be provided intermittently, as shown in the sectional view at the top of Fig. 4.7(a).

2. The lacing provides lateral support for the component column segments at each connecting point. For example, in Fig. 4.7(a), the l_0/r_0 of each individual angle, between support points, must be less than the overall l/r of the whole member. The r_0 of the individual angle should be the

Fig. 4.6 W 14 steel column in substructure of a multistory building. (Courtesy American Institute of Steel Construction.)

minimum value, as tabulated in the AISCM for the inclined z-z axis (AISCS, Sec. 1.18.2.6).

3. The lacing acts as a web replacement to resist shear and provides for the corresponding transfer of longitudinal variations of stress in the component longitudinal elements. In a centrally loaded column, shear force arises from accidental end eccentricity of the load and from curvature of the member under load. The shear force provides the basis for lacing bar design, and under the AISCS, Sec. 1.18.2.6, the resultant shear for design is taken as 2% of the axial load in the member. Lacing bars may be designed as secon-

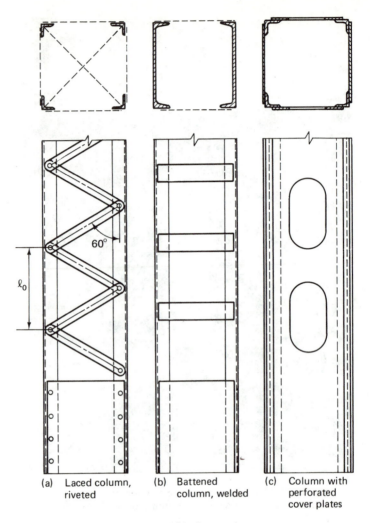

Fig. 4.7 Types of built-up columns.

dary members, but must act either in tension or compression and be designed for both loading conditions. Complete design rules for built-up members, including laced members and their end tie plates, are provided in Sec. 1.18, AISCS.

Battened columns [Fig. 4.7(b)] are not covered by the AISCS, but are used occasionally. They present the same design problems as do laced columns, but in comparison with the truss action of the lacing members, the battens resist shear by the less effective and more complex continuous-frame action.

Tie plates at the ends of both laced and battened members are particularly important to distribute the applied end loads. In weakly battened columns they may add materially to the overall column strength. For a more complete coverage of design of both laced and battened columns, reference should be made to Chapter 12 of the SSRC *Guide* (3rd edition).

Columns with perforated cover plates [Fig. 4.7(c)] are used chiefly in bridge construction (see also Fig. 2.6). The net section of such plates may be included in the column area, and they resist shear more effectively than either the laced or battened column. The perforations are provided primarily for drainage in exposed location and to provide access for cleaning and painting the interior surfaces. Simple design rules for such members are provided in all modern bridge specifications.

4.5 WIDTH/THICKNESS RATIOS

Width/thickness limitations are established to ensure that overall column buckling rather than local buckling governs the allowable design stress. When the limitations are not exceeded, the full cross section of the column may be considered to be effective. Limits on width/thickness ratios have been discussed in Sec. 4.4 for the case of the angle and tee section.

Limitations are as specified in the AISCS, Sec. 1.9. Width/thickness limits are established under two broad categories, *unstiffened elements* and *stiffened elements*, as defined in AISCS, Sec. 1.9, and illustrated herein in Fig. 4.8. For equal width/thickness ratios, a stiffened element is much more effective than an unstiffened element, and much greater ratio limits are allowed for the stiffened element. As the yield stress increases, a more stocky element (smaller width/thickness ratio) is required to prevent premature local buckling under the increased allowable stress.

When a thin-walled member does double duty both as a column and partition, it may be desirable to exceed the width/thickness limits. Such members may be used provided a "reduced effective width" and/or reduced allowable stresses are employed, as covered in Appendix C of the AISCS or in the AISI specifications. Reference also may be made to Chapter 9 of the SSRC *Guide* (3rd edition).

4.6 COLUMN BASE PLATES AND SPLICES

Columns on footings must be provided with base plates to distribute load to the masonry within the allowable bearing capacity of the concrete. Permissible masonry bearing pressures are specified in Sec. 1.5.5 of the AISCS. For very

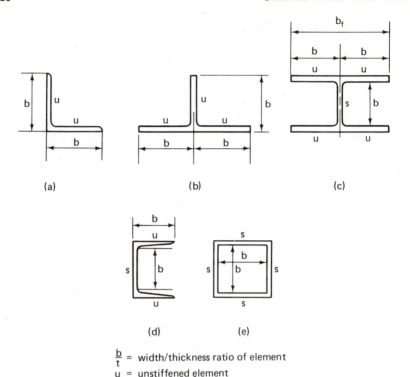

$\dfrac{b}{t}$ = width/thickness ratio of element

u = unstiffened element

s = stiffened element

Fig. 4.8 Stiffened and unstiffened elements of structural shapes as defined for AISCS Sec. 1.9 width/thickness limitations.

heavy columns in tall buildings, individual base plates may not be sufficient, and a grillage system may be required. For information on base-plate design, including recommended design procedures, reference should be made to the AISCM.

4.7 ALLOWABLE COMPRESSION STRESS

Insofar as general buckling under axial load governs the design selection of a steel column, the allowable stress at working load is defined by the AISCS, Sec. 1.5.1.3, and is based on the column-strength curves, Eqs. (4.4) and (4.3), divided by an appropriate factor of safety (*FS*).

   (a) When $Kl/r < C_c$, apply the CRC basic column-strength curve, Eq. (4.4); then the allowable compression stress is

$$F_a = \left[1 - \frac{(Kl/r)^2}{2C_c^2}\right]\frac{F_y}{FS} \tag{4.7}$$

where

$$FS = \frac{5}{3} + \frac{3}{8C_c}\left(\frac{Kl}{r}\right) - \frac{1}{8C_c^3}\left(\frac{Kl}{r}\right)^3 \tag{4.8}$$

(b) When $C_c < Kl/r < 200$, apply the Euler formula, Eq. (4.3); then

$$F_a = \frac{\pi^2 E}{FS(Kl/r)^2} \tag{4.9}$$

where

$$FS = \tfrac{23}{12} = 1.92 \tag{4.10}$$

It is noted that the allowable compression stress is independent of the yield point when $Kl/r > C_c$.

At Kl/r of zero, the factor of safety of 1.67 in compression is the same as for tension. With increasing Kl/r, the factor of safety increases toward 1.92. This allows for uncertainties such as unavoidable eccentricity, residual stress, crookedness, and, for very slender columns, the increased sensitivity to uncertainties in the evaluation of effective length. The foregoing allowable stresses are limited to members that meet the width/thickness requirements of AISCS, Sec. 1.9.

For bracing and secondary members, when l/r is between 120 and 200, the allowable compression stress F_{as} is

$$F_{as} = \frac{F_a}{1.6 - (l/200r)} \tag{4.11}$$

Allowable compression stress for steels having yield stresses of 36 and 50 ksi are tabulated in Tables 3-36 and 3-50 of Appendix A of the AISCS. For all values of F_y other than 36 and 50 ksi, Appendix A of AISCS provides Tables 4 and 5, which simplify the determination of F_a by Eqs. (4.7) and (4.8). Table 4 provides values of C_a as a function of Kl/rC_c for use in Eq. (4.12):

$$F_a = C_a F_y \tag{4.12}$$

The solution of a column design problem, using either calculator or desk computer, illustrates very well the basic difference between a problem of *analysis* of a given structure and the problem of *design*, for which the rules (specifications) are given and the structure is the end product. In column design the allowable stress is a function of the slenderness ratio and is therefore generally unknown in advance of a trial design selection. Exceptions to this statement are represented by the tube, of a given diameter, and the built-up column of four angles. In these cases the suitable general dimensions of the cross section can usually be established in advance, thus pinpointing the slenderness ratio, and the final design is arrived at by simply changing the wall or angle thickness to provide the required area that is equal to the load divided by the allowable stress. Similarly, in heavy tier building construction, using W 14 shapes, the radius of gyration of the cross section will be known approximately in advance. But in many cases it may be necessary to simply "guess" at some preliminary trial selection and arrive at the final design by means of one or two additional trial selections. One may either guess the radius of gyration, or, more directly, guess the allowable stress. The analysis of such a trial selection will quickly lead to a much better trial. Most of the examples presented herein involve the check of a final trial selection—in each case the reader should ask himself: How best could this trial have been arrived at?

Flowchart 4.1 systematizes the use of the AISCS for the selection of allowable column stress under axial load. It will also be used as an adjunct to Chapter 5 for the design of beam-columns, as the value of F_a enters into the AISCS interaction formula for beam-column design.

4.8 FLOWCHARTS

Flowchart 4.1

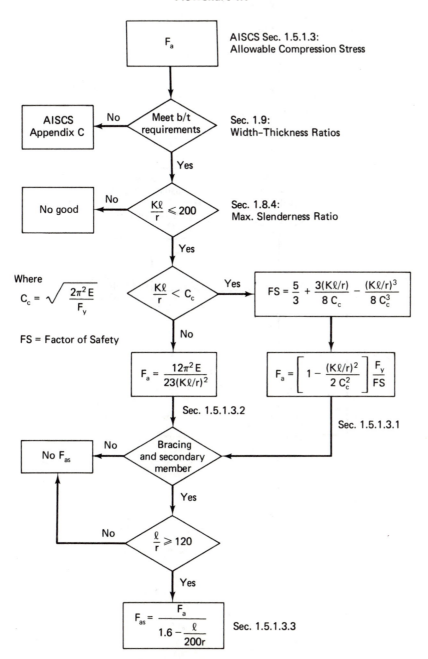

F_a

AISCS Sec. 1.5.1.3:
Allowable Compression Stress

Meet b/t requirements — No → AISCS Appendix C

Sec. 1.9:
Width-Thickness Ratios

Yes

$\frac{K\ell}{r} \leqslant 200$ — No → No good

Sec. 1.8.4:
Max. Slenderness Ratio

Yes

Where

$$C_c = \sqrt{\frac{2\pi^2 E}{F_y}}$$

FS = Factor of Safety

$\frac{K\ell}{r} < C_c$ — Yes →

$$FS = \frac{5}{3} + \frac{3(K\ell/r)}{8\,C_c} - \frac{(K\ell/r)^3}{8\,C_c^3}$$

No

$$F_a = \frac{12\pi^2 E}{23(K\ell/r)^2}$$

$$F_a = \left[1 - \frac{(K\ell/r)^2}{2\,C_c^2} \right] \frac{F_y}{FS}$$

Sec. 1.5.1.3.2

Sec. 1.5.1.3.1

Bracing and secondary member — No → No F_{as}

Yes

$\frac{\ell}{r} \geqslant 120$ — No →

Yes

$$F_{as} = \frac{F_a}{1.6 - \dfrac{\ell}{200r}}$$

Sec. 1.5.1.3.3

Flowchart 4.2

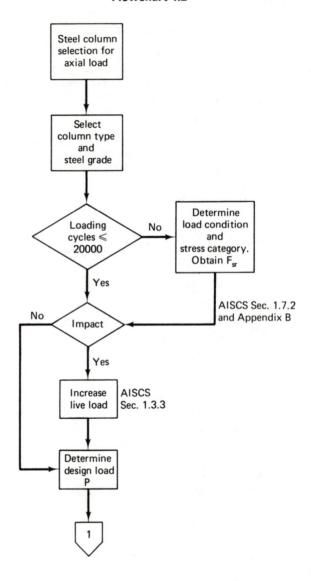

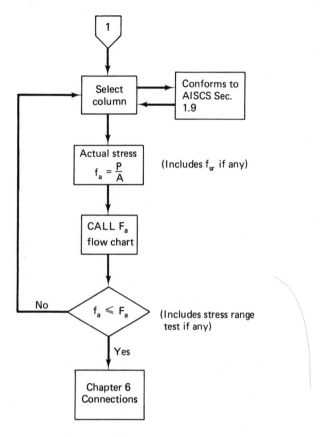

4.9 ILLUSTRATIVE EXAMPLES

Example 4.1

A portion of a trussed TV antenna tower, as shown, has main longitudinal elements that carry an axial load P (270 kips) and are laterally braced at 6-ft intervals. No advantage will be taken of rotation restraint provided by lateral bracing at

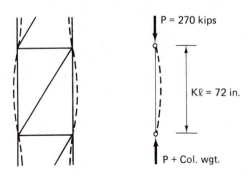

ends. The dashed line indicates a natural buckling mode of the main compression elements at failure. Select a solid round bar for the main compression elements to satisfy the AISCS. Use A36 steel.

Solution

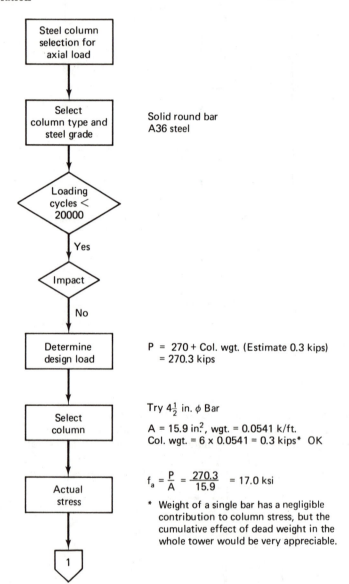

Solid round bar
A36 steel

$P = 270 + $ Col. wgt. (Estimate 0.3 kips)
$\quad = 270.3$ kips

Try $4\frac{1}{2}$ in. ϕ Bar

$A = 15.9$ in.^{2}, wgt. = 0.0541 k/ft.
Col. wgt. = $6 \times 0.0541 = 0.3$ kips*   OK

$$f_a = \frac{P}{A} = \frac{270.3}{15.9} = 17.0 \text{ ksi}$$

* Weight of a single bar has a negligible contribution to column stress, but the cumulative effect of dead weight in the whole tower would be very appreciable.

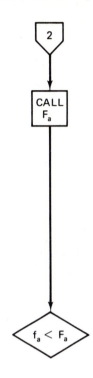

See F_a flow chart

Effective length factor K = 1.0 (see Table 4.1)

Radius of gyration $r = \dfrac{d}{4} = \dfrac{4.5}{4} = 1.125$ in.

Slenderness ratio $\dfrac{K\ell}{r} = \dfrac{1.0(6 \times 12)}{1.125} = 64.0$

$C_c = \sqrt{2\pi^2 E/F_y} = \sqrt{2\pi^2 \times 29000/36} = 126.1$

(see AISCS Table 5, Appendix A)

$\dfrac{K\ell}{r} < C_c$, then

$FS = \dfrac{5}{3} + \dfrac{3 \times 64}{8 \times 126.1} - \dfrac{64^3}{8 \times 126.1^3}$

$= 1.844$

$F_a = [1 - \dfrac{64^2}{2 \times 126.1^2}] \dfrac{36}{1.844} = 17.04$ ksi

(see AISCS Table 3-36, Appendix A)

Stress test OK

USE a Bar $4\frac{1}{2}$ ϕ

Note: In computer aided design, formulas for C_c, F_a, etc. would be part of the program. In manual design, maximum use would be made of charts and tables, as referred to above. Calculation details as given above are to provide a complete illustration of specification requirements and are not typical of actual more abbreviated design computations.

Example 4.2

For an unbraced and effectively hinged-end length of 8-ft, design a steel pipe column for an axial load of 50 kips, using A36 steel, AISCS, and limiting the selection to standard pipe sizes.

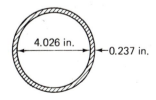

4.026 in. ←0.237 in.

Solution

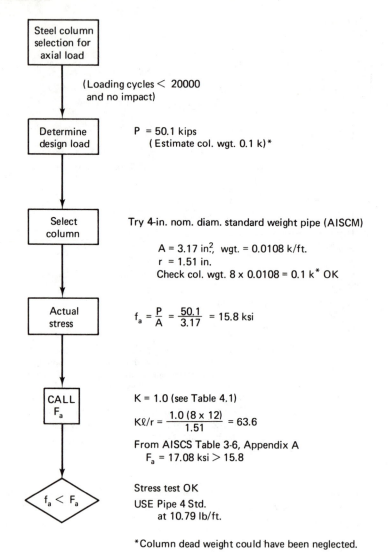

Steel column selection for axial load

(Loading cycles $<$ 20000 and no impact)

Determine design load

P = 50.1 kips
(Estimate col. wgt. 0.1 k)*

Select column

Try 4-in. nom. diam. standard weight pipe (AISCM)

A = 3.17 in.², wgt. = 0.0108 k/ft.
r = 1.51 in.
Check col. wgt. 8 x 0.0108 = 0.1 k* OK

Actual stress

$f_a = \dfrac{P}{A} = \dfrac{50.1}{3.17} = 15.8$ ksi

CALL F_a

K = 1.0 (see Table 4.1)

$K\ell/r = \dfrac{1.0\,(8 \times 12)}{1.51} = 63.6$

From AISCS Table 3-6, Appendix A
F_a = 17.08 ksi $>$ 15.8

$f_a < F_a$

Stress test OK
USE Pipe 4 Std.
at 10.79 lb/ft.

*Column dead weight could have been neglected.

Example 4.3

Design a square structural tube for the same conditions as in Example 4.2.

Solution

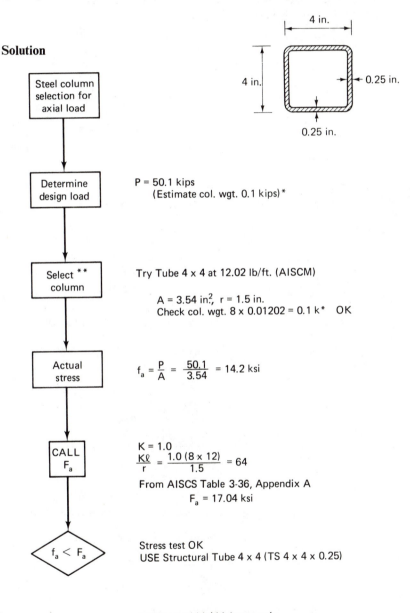

P = 50.1 kips
   (Estimate col. wgt. 0.1 kips)*

Try Tube 4 x 4 at 12.02 lb/ft. (AISCM)

   A = 3.54 in.^{2}, r = 1.5 in.
   Check col. wgt. 8 x 0.01202 = 0.1 k*   OK

$$f_a = \frac{P}{A} = \frac{50.1}{3.54} = 14.2 \text{ ksi}$$

K = 1.0
$$\frac{K\ell}{r} = \frac{1.0\,(8 \times 12)}{1.5} = 64$$
From AISCS Table 3-36, Appendix A
      F_a = 17.04 ksi

Stress test OK
USE Structural Tube 4 x 4 (TS 4 x 4 x 0.25)

**Check width/thickness ratio
   (AISCS Sec. 1.9.2.2)
   $$\frac{b}{t} = \frac{4}{0.25} = 16 < 39.7 \text{ OK}$$

* Dead weight could have been neglected.

Example 4.4

Design a double-angle strut for the same conditions as in Example 4.2 with the added stipulation that special attention must be given to end connections to bring the axial load uniformly into both legs at each end.

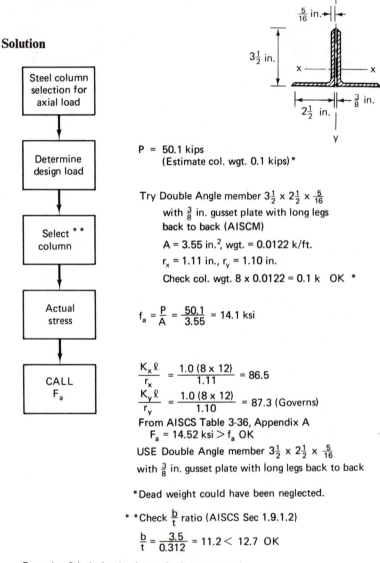

Solution

Steel column selection for axial load

↓

Determine design load

P = 50.1 kips
(Estimate col. wgt. 0.1 kips)*

↓

Select** column

Try Double Angle member $3\frac{1}{2} \times 2\frac{1}{2} \times \frac{5}{16}$
with $\frac{3}{8}$ in. gusset plate with long legs
back to back (AISCM)

A = 3.55 in.², wgt. = 0.0122 k/ft.

r_x = 1.11 in., r_y = 1.10 in.

Check col. wgt. 8 x 0.0122 = 0.1 k   OK *

↓

Actual stress

$f_a = \dfrac{P}{A} = \dfrac{50.1}{3.55} = 14.1$ ksi

↓

CALL F_a

$\dfrac{K_x \ell}{r_x} = \dfrac{1.0\,(8 \times 12)}{1.11} = 86.5$

$\dfrac{K_y \ell}{r_y} = \dfrac{1.0\,(8 \times 12)}{1.10} = 87.3$ (Governs)

From AISCS Table 3-36, Appendix A
  F_a = 14.52 ksi > f_a OK

USE Double Angle member $3\frac{1}{2} \times 2\frac{1}{2} \times \frac{5}{16}$
with $\frac{3}{8}$ in. gusset plate with long legs back to back

*Dead weight could have been neglected.

**Check $\dfrac{b}{t}$ ratio (AISCS Sec 1.9.1.2)

$\dfrac{b}{t} = \dfrac{3.5}{0.312} = 11.2 < 12.7$ OK

Remark: Stitch riveting is required to prevent the premature buckling of a single angle, then

$\dfrac{K\ell}{r} < 87.3$, where K = 1.0, r = r_z = 0.54 in.

$\ell < \dfrac{87.3r}{K} = \dfrac{87.3 \times 0.54}{1.0} = 47.2$ in.

USE spacing of 3 ft 6 in. c. to c. stitch rivets.

Example 4.5

Design a structural tee column for the same conditions as in Example 4.2.

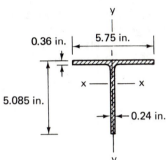

Solution

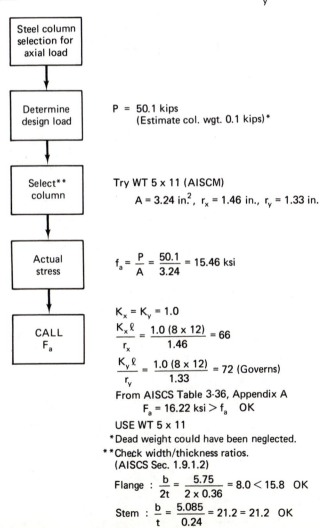

Steel column selection for axial load

Determine design load

$P = 50.1$ kips
  (Estimate col. wgt. 0.1 kips)*

Select** column

Try WT 5 x 11 (AISCM)
  $A = 3.24$ in.^{2}, $r_x = 1.46$ in., $r_y = 1.33$ in.

Actual stress

$f_a = \dfrac{P}{A} = \dfrac{50.1}{3.24} = 15.46$ ksi

CALL F_a

$K_x = K_y = 1.0$

$\dfrac{K_x \ell}{r_x} = \dfrac{1.0 \, (8 \times 12)}{1.46} = 66$

$\dfrac{K_y \ell}{r_y} = \dfrac{1.0 \, (8 \times 12)}{1.33} = 72$ (Governs)

From AISCS Table 3-36, Appendix A
  $F_a = 16.22$ ksi $> f_a$ OK

USE WT 5 x 11

  *Dead weight could have been neglected.
**Check width/thickness ratios.
  (AISCS Sec. 1.9.1.2)

Flange : $\dfrac{b}{2t} = \dfrac{5.75}{2 \times 0.36} = 8.0 < 15.8$ OK

Stem : $\dfrac{b}{t} = \dfrac{5.085}{0.24} = 21.2 = 21.2$ OK

Example 4.6

Design a wide-flange (W) column 20 ft in length to support an axial load of 410 kips in the interior of a building. The column base is rigidly fixed to the footing and the top of the column is rigidly framed to very stiff girders. Assume that bracing is provided to prevent sidesway in the weak deflection plane of the column, but the sidesway in the strong plane is not prevented. Select an economical W shape to satisfy the AISCS. Use A36 steel.

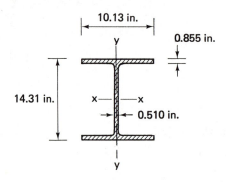

Solution

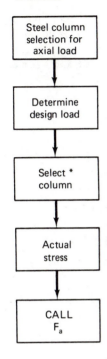

P = 411.6 kips
  (Estimate col. wgt. 1.6 kips)

Try W 14 x 82 (AISCM)
  A = 24.1 in.^{2}, r_x = 6.05 in., r_y = 2.48 in.
  Check col. wgt. 20 x 0.082 = 1.6 k   OK

$$f_a = \frac{P}{A} = \frac{411.6}{24.1} = 17.08 \text{ ksi}$$

K_x = 1.2, K_y = 0.65 (see Table 4.1)

$$\frac{K_x \ell}{r_x} = \frac{1.2\,(20 \times 12)}{6.05} = 47.6$$

$$\frac{K_y \ell}{r_y} = \frac{0.65\,(20 \times 12)}{2.48} = 62.9 \text{ (Governs)}$$
(Governing K = 0.65, r = 2.48 in.)
From AISCS Table 3-36, Appendix A

  F_a = 17.15 ksi $>$ f_a OK

USE W 14 x 82

* Check width/thickness ratios
  (AISCS Sec. 1.9.1. and 1.9.2 and Table 6, Appendix A)

  Web: $\dfrac{h}{t} = \dfrac{14.31}{0.510} = 28.1 < 42.2$ OK

  Flange: $\dfrac{b}{t} = \dfrac{10.13/2}{0.855} = 5.92 < 15.8$ OK

Example 4.7†

Same as Example 4.6, except the top of the column is rigidly framed to two-way floor beams, as shown in the figure. Determine the axial load capacity of the selected W 14 × 82 column in the previous example. Assume that all stories are spaced at 20 ft and adjacent stories have the same column section. Use AISCS and A36 steel.

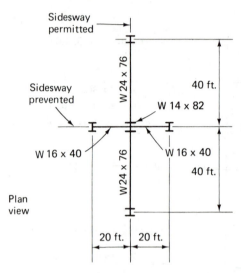

Member properties:

W 14 × 82 column: $A = 24.1$ in.^{2} $I_x = 882$ in.^{4} $I_y = 148$ in.^{4}
                         $r_x = 6.05$ in. $r_y = 2.48$ in.

W 24 × 76 beam: $I_x = 2100$ in.^{4}

W 16 × 40 beam: $I_x = 518$ in.^{4}

Solution

Determine the governing slenderness ratio.

Case 1

Major x axis (sidesway permitted): at the column top,

$$G_A = \frac{\sum I_c/L_c}{\sum I_g/L_g}$$

$$= \frac{(882/20) + (882/20)}{(2100/40) + (2100/40)} = 0.840$$

†See AISCM for an explicit illustration of the use of the alignment chart prior to study of this example.

At the column bottom $G_B = 1.0$ (AISCS Commentary) for column base rigidly attached to a footing.

Apply Fig. 4.4(b) for $G_A = 0.840$, $G_B = 1.0$; then

$$K_x = 1.29$$

$$\frac{K_x l}{r_x} = \frac{1.29(20 \times 12)}{6.05} = 51.2$$

Case 2

Minor y axis (sidesway prevented)†: at the column top,

$$G_A = \frac{(148/20) + (148/20)}{(518/20) + (518/20)} = 0.286$$

At the column bottom, $G_B = 1.0$. Apply Fig. 4.4(a) for $G_A = 0.286$, $G_B = 1.0$; then

$$K_y = 0.695$$

$$\frac{K_y l}{r_y} = \frac{0.695(20 \times 12)}{2.48} = 67.3$$

Therefore, the slenderness ratio is governed by case 2, for which the buckling will occur in the column about the minor y axis:

$$\frac{Kl}{r} = 67.3$$

Therefore,

$$F_a = 16.71 \text{ ksi} \qquad \text{(from AISCS, Table 3-36, Appendix A)}$$

The axial load capacity of the W 14 × 82 column is

$$P = F_a A = 16.71 \times 24.1 = 403 \text{ kips}$$

Further study of problems relating to the design of columns in frames and the interaction between framed columns is provided in Section 8.4.

---

†Since the AISCS Commentary does not include the alignment chart for "sidesway prevented," one might conservatively take $K_y = 0.8$, as recommended in Table 4.1 for the case with fixed base and hinged top. In that case

$$\frac{K_y l}{r_y} = \frac{0.80 \times 240}{2.48} = 77.4$$

$$F_a = 15.65 \text{ ksi}$$

$$P = 15.65 \times 24.1 = 377 \text{ kips}$$

PROBLEMS

4.1. Determine the diameter of a solid round A36 steel bar, 8 ft in length, that will support an axial force of 160 kips. Choose an available size. Lateral support is assumed at each end. The ends are assumed to be hinged; hence the effective column length is the same as the actual length. Redesign, using A514 steel, with $F_y = 90$ ksi. Compare the weights.

4.2. Select standard A36 steel shapes having an area as near as possible to that of the requirement determined for Problem 4.1 and compare the column load capacities for the following:
   a. A W 8 section.
   b. A single angle with equal legs.
   c. A double angle with long legs spaced $\frac{3}{8}$ in. apart.
   d. A structural tee.
   e. A round pipe.†
   f. A square structural tube.†

4.3. Assume for the purposes of this problem that the total weight of a 30-story building and its contents (vertical live load) averages 160 lb/ft² of floor area per floor. For a floor plan 90 ft by 180 ft, encompassing 40 columns, make a preliminary first-floor trial-column-size selection for an effective column length of 14 ft. There are four rows spaced 30 ft apart, of 10 columns each. Columns are assumed to be axially loaded and the choice can be made with the aid of any available column load tables—but the actual capacity of the column for axial load should be checked by the AISCS formula. Assume each column to be loaded by floor areas tributary to it.

4.4. Rework Problem 4.3 on the assumption that the columns will be braced laterally in their weak direction.

4.5. A WT 7 × 21.5 is welded to a PL $\frac{7}{8}$ × 14 as shown. The overall length is 24 ft and the column is fixed at the base, hinged at the top. Use recommended effective length factor from Table 4.1. Determine the allowable column load for (a) A36 steel, and (b) a steel with $F_y = 50$ ksi.

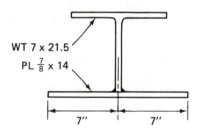

WT 7 x 21.5
PL $\frac{7}{8}$ x 14

7"   7"

†Use the largest available outside dimension for the given area. (*Note:* AISCS, Appendix A, Table 3-36, may be used.)

4.6. A hinged end column 32 ft in length is supported in its weak direction only, at the midpoint, in which case it may be assumed to behave as if the effective length for buckling about the y–y axis is 16 ft. Select a section to carry a load of 110 kips, using A36 steel. Provide as nearly equal slenderness ratios as possible about the two axes.

4.7. (See the illustration.) Considering the column cross section, length, local details, and bracing, as shown, what is the permissible load P utilizing A36 steel? See AISCS, Sec. 1.18.2.6, for design assumptions and select an adequate size for the lacing bars, which may be assumed to have adequate welded end connections and which need to be selected on the basis of adequate compressive strength.

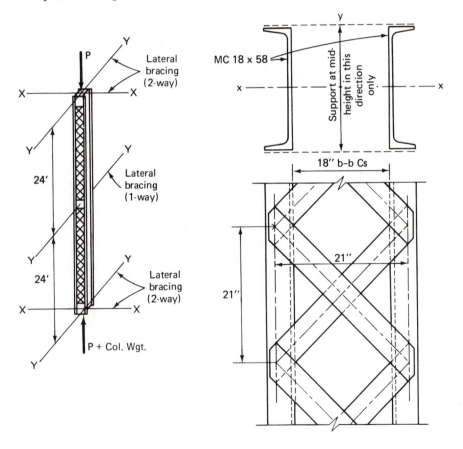

4.8. Design a column 48 ft in length to carry a load of 500 kips without any intermediate lateral support. Use a cross section consisting of four corner angles with double lacing similar to that used in Problem 4.7 on all four sides

(See the sketch of the cross section.) Design all details, including size of lacing, except for welded connections between lacing and angles. Follow requirements of AISCS, Sec. 1.18.2.6, and use a steel with $F_y = 50$ ksi.

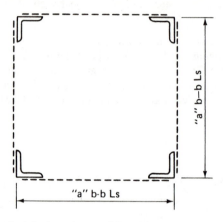

4.9. Rework Example 4.6, changing steel from A36 to $F_y = 50$ ksi.

4.10. Rework Example 4.7, changing steel from A36 to $F_y = 50$ ksi

5

COLUMNS UNDER COMBINED STRESS†

5.1 INTRODUCTION

In Chapter 3 we treated steel beam design and in Chapter 4 the design of axially loaded columns. In actual structures, most columns, in addition to axial load, must support lateral loads and/or transmit moments between their ends, and are thus subjected to combined stress due both to axial load and moment. Such members are termed beam-columns. The end moments may be caused by continuous frame action and/or by the effective eccentricity of the longitudinal loads. For example, columns in a tall building frame, in addition to the live and dead loads of the structure above any given level, must transmit bending moments that result from wind load or lateral inertia forces due to earthquake. They must also resist the end moments introduced through the continuous frame action of the loaded adjacent connecting beams. In building frames the moments induced by the sway deflection of the structure must also be included.

When a column is part of a frame, the ideal solution would be to determine the strength of the complete structure and use this as a basis for design. There is a trend toward such a design procedure, but at this time the traditional method of isolating the individual member as a basis for design prevails. The design may then proceed along one of the following three paths:

†The application of AISCS, Sec. 1.6.2, to the tension member under combined stress is also illustrated in Example 5.5.

1. The load at which the maximum stress reaches the yield stress is determined and that load is divided by a factor of safety to give an allowable load. Moment due to deflection is included.

2. The AISCS interaction formula may be used. This provides for an empirical transition in the member selection, in accordance with AISCS requirements, from the beam (as the column load approaches zero) to the axially loaded column (as the bending moment approaches zero).

3. The AISCS plastic-design procedure, also using an interaction formula similar to that mentioned in item 2, will be used if this alternative (plastic-design) method is adopted for continuous steel frame design. The principal difference is that under item 2 the moments are determined by elastic frame analysis (such as the moment-distribution method), whereas in plastic design the bending moments are distributed as they would be when the frame is at the verge of collapse.

In Chapter 8 the design of columns as part of a one-story frame is treated by both the allowable-stress and plastic-design procedures.

5.2   ALLOWABLE-STRESS DESIGN

A very short and stocky beam-column of A36 steel in which Kl/r is less than 15,† may be designed very simply and quite conservatively on the basis of full maximum allowable stress. In such a short member the additional bending moment due to column curvature may be neglected. The designation "stocky" is important, because even a short member is subject to failure by local buckling and must meet the width/thickness requirements for compression members. The maximum stress is equal to the sum of the average stress due to axial load plus the compressive stress added by bending moment:

$$f_{max} = \frac{P}{A} + \frac{M_x c_x}{I_x} \tag{5.1}$$

A more direct selection of a short beam-column is made possible by rewriting Eq. (5.1) along the following lines. I_x may be replaced by Ar_x^2 and, as shown in Fig. 5.1, the moment M may be replaced by its static equivalent, Pe_x; thus

$$f_{max} = \frac{P}{A}\left(1 + \frac{e_x c_x}{r_x^2}\right) \tag{5.2a}$$

$$\frac{P}{A} = \frac{f_{max}}{1 + (e_x c_x/r_x^2)} \tag{5.2b}$$

†For $F_y > 36$ ksi, use $90/\sqrt{F_y}$ in place of 15.

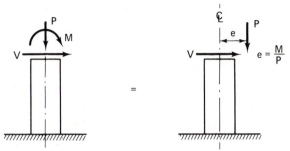

Fig. 5.1 Equivalent loads applied to a short beam-column.

Thus, by means of Eq. (5.2b), the average allowable column stress may be obtained very simply by substituting the full allowable stress for a zero-length column $(0.6F_y)$ in place of f_{max}. For the stipulated limit on l/r, the allowable axial column stress will be no more than about 3% less than this value. If the section is "compact," the bending part of the stress would have a higher allowable value of $0.66F_y$, a feature that is introduced by the interaction formula to be discussed in the next section.

If there are components of moment about two axes of the cross section, Eq. (5.2b) may be modified to the following:

$$\frac{P}{A} = \frac{f_{max}}{1 + (e_x c_x/r_x^2) + (e_y c_y/r_y^2)} \tag{5.3}$$

In applying either Eq. (5.2b) or Eq. (5.3) to design, values of c and r about each axis may be approximated by reference to the design information in the AISCM for a particular series of sections. A trial selection is then made on the basis of the required P/A. If satisfactory, the trial values of c and r may then be replaced by the actual values for the selected section and the design rechecked.

Although Eq. (5.3) may be further modified† to adapt it to longer columns by including the effect of deflection, attention is now turned to the AISCS interaction formula procedure, as currently used in design of beam-columns.

5.3 DESIGN BY USE OF INTERACTION FORMULAS

Using AISCS notation, Eq. (5.1) may be written

$$f_{max} = f_a + f_{bx}$$

or, if there are moments about both axes of the cross section,

$$f_{max} = f_a + f_{bx} + f_{by}$$

†See *The USS Steel Design Manual* (Ref. 1.7) for this procedure.

Dividing both sides of the foregoing by f_{max},

$$1 = \frac{f_a}{f_{max}} + \frac{f_{bx}}{f_{max}} + \frac{f_{by}}{f_{max}} \qquad (5.4)$$

Equation (5.4) is now in the form termed an interaction formula, but for design application it can be improved by introducing the allowable stress applicable to each of the three terms on the right-hand side of Eq. (5.4) in place of f_{max}. Then if any two of the terms on the right side become zero, the correct specified allowable stress is approached in the limit as the beam-column becomes an axially loaded column or a beam bent about either of the two axes. With such a modification Eq. (5.4) becomes the AISCS interaction equation for design under combined stress for the case where f_a/F_a is less than 0.15:

$$\frac{f_a}{F_a} + \frac{f_{bx}}{F_{bx}} + \frac{f_{by}}{F_{by}} \leq 1.0 \qquad \text{[AISCS (1.6-2)]}$$

When f_a/F_a is greater than 0.15, additional bending moments due to column curvature deflection ($P\delta_x$ and $P\delta_y$) may contribute significantly to the combined stress (see Fig. 2.2). The additional bending moment may be approximated by multiplying both f_{bx} and f_{by} by an *amplification factor*,

$$\frac{C_m}{1 - (f_a/F'_e)}$$

where F'_e is the Euler buckling stress [Eq. (4.3)] divided by $\frac{23}{12}$, or 1.92, which is the AISCS factor of safety for the very long column with Kl/r greater than C_c. C_m is a correction factor to the standard amplification, usually about 1, which adjusts for variations in the distribution of bending moment along the member, as covered by Sec. 1.6.1 of the AISCS. At this point the AISCS Commentary should be studied relative to AISCS, Sec. 1.6.1.

With the introduction of the amplification factors, AISCS Eq. (1.6-2) is modified so as to be applicable to the case when f_a/F_a is greater than 0.15, as follows:

$$\frac{f_a}{F_a} + \frac{C_{mx}f_{bx}}{(1 - f_a/F'_{ex})F_{bx}} + \frac{C_{my}f_{by}}{(1 - f_a/F'_{ey})F_{by}} \leq 1.0 \qquad \text{[AISCS (1.6-1a]}$$

and, in addition, at points braced in the plane of bending,

$$\frac{f_a}{0.6F_y} + \frac{f_{bx}}{F_{bx}} + \frac{f_{by}}{F_{by}} \leq 1.0 \qquad \text{[AISCS (1.6-1b)]}$$

where the subscripts x and y indicate the axis of bending about which a particular stress or design property applies, and

F_a = allowable axial compression stress for axial force alone

F_b = allowable bending stress for bending moment alone

$$F'_e = \frac{12\pi^2 E}{23(Kl_b/r_b)^2} \qquad \text{(see AISCS, Table 2)}$$

where l_b is the actual unbraced length in the plane of bending and r_b is the corresponding radius of gyration. K is the effective length factor in the plane of bending. As in the case of F_a, F_b, and $0.6F_y$, F'_e may be increased 33.3% for wind and seismic loads in accordance with Sec. 1.5.6.

f_a = actual axial compression stress

f_b = actual maximum fiber compressive bending stress

C_m is a coefficient defined in Sec. 1.6.1 of AISCS as follows:

1. For compression members in frames subject to sidesway, $C_m = 0.85$.
2. For restrained compression members in frames braced against sidesway and not subject to transverse loading between their supports in the plane of bending,

$$C_m = 0.6 - 0.4\frac{M_1}{M_2}$$

but not less than 0.4, where M_1/M_2 is the ratio of the smaller to larger moments at the ends of that portion of the member unbraced in the plane of bending under consideration. M_1/M_2 is positive when the member is bent in reverse curvature and negative when it is bent in single curvature.

3. For compression members in frames braced against sidesway in the plane of loading and subjected to transverse loading between their supports, the value of C_m may be determined by rational analysis. However, in lieu of such analysis the following values may be used: (a) for members whose ends are restrained, $C_m = 0.85$; (b) for members whose ends are unrestrained, $C_m = 1.0$.

5.4 EQUIVALENT AXIAL COMPRESSION LOAD

Both the preliminary selection and final design check of a beam-column, proportioned by the AISCS interaction formulas, may be expedited by conversion to an equivalent axial load, thereby making use of the tabulated loads provided in the AISCM. Any one of the AISCS interaction formulas (1.6-1a), (1.6-1b), or (1.6-2) can be generalized as follows[†]:

$$\alpha_a\frac{f_a}{F_a} + \alpha_{bx}\frac{f_{bx}}{F_{bx}} + \alpha_{by}\frac{f_{by}}{F_{by}} \leq 1.0$$

Multiplying both sides of the inequality by AF_a,

$$\alpha_a Af_a + \alpha_{bx}\frac{f_{bx}}{F_{bx}}AF_a + \alpha_{by}\frac{f_{by}}{F_{by}}AF_a \leq AF_a$$

[†]α_a, α_{bx}, and α_{by} are simply coefficients that represent a collection of terms in any one of the three AISCS formulas. In the very short strut each $\alpha = 1$ [see Eq. (5.4)].

AF_a would be the allowable load if the column were axially loaded. Thus the sum of the three terms on the left of the foregoing inequality can be thought of as an equivalent axial load and used as an entry into the AISCM column-selection tables. The use of the AISCM is further facilitated by rewriting the previous equation in terms of M_x and M_y, and by introducing coefficients B_x and B_y, which are also tabulated in the AISCM:

$$Af_{bx} = \frac{M_xA}{S_x} = M_xB_x \qquad Af_{by} = \frac{M_yA}{S_y} = M_yB_y$$

Noting that Af_a is the actual column load, the equivalent load equation, in general terms, becomes

$$P_{eq} = \alpha_a P + \alpha_{bx}\frac{F_a}{F_{bx}}B_xM_x + \alpha_{by}\frac{F_a}{F_{by}}B_yM_y \qquad (5.5)$$

We may now rewrite AISCS Eqs. (1.6-1a), (1.6-1b), and (1.6-2) as follows: Modified AISCS Eq. (1.6-1a), for $f_a/F_a > 0.15$:

$$P_{eq} = P + B_xM_xC_{mx}\frac{F_a}{F_{bx}}\frac{1}{1-(P/AF'_{ex})} + B_yM_yC_{my}\frac{F_a}{F_{by}}\frac{1}{1-(P/AF'_{ey})}$$

Although the foregoing may be used as it stands, the AISCM further simplifies the amplification portions by the tabulation of additional section coefficients a_x and a_y. Introducing the formula for F'_e,

$$AF'_e = \frac{12\pi^2EAr^2}{23(Kl)^2}$$

we let a (subscripts x and y temporarily omitted) be

$$a = \frac{12\pi^2EAr^2}{23}$$

With the further simplification of AISCS Eq. (1.6-1a), all three modified formulas are now summarized:

$$P_{eq} = \text{required tabular load}$$

$$= P + B_xM_xC_{mx}\left(\frac{F_a}{F_{bx}}\right)\left(\frac{a_x}{a_x - P(Kl)^2}\right)$$

$$+ B_yM_yC_{my}\left(\frac{F_a}{F_{by}}\right)\left(\frac{a_y}{a_y - P(Kl)^2}\right)$$

modified formula (1.6-1a)

$$P_{eq} = \text{required tabular load}$$

$$= P\left(\frac{F_a}{0.6F_y}\right) + B_xM_x\left(\frac{F_a}{F_{bx}}\right) + B_yM_y\left(\frac{F_a}{F_{by}}\right)$$

modified formula (1.6-1b)

When $f_a/F_a \leq 0.15$,

$$P_{eq} = \text{required tabular load}$$

$$= P + B_x M_x \left(\frac{F_a}{F_{bx}}\right) + B_y M_y \left(\frac{F_a}{F_{by}}\right)$$

modified formula (1.6-2)

In Formula (1.6-1a), for the term $(Kl)^2$, K is the effective length factor and l is the actual unbraced length in the plane of bending. Values for the components a_x and a_y, equal to $0.149 \times 10^6 Ar_x^2$ and $0.149 \times 10^6 Ar_y^2$, respectively, are listed at the bottom of the load tables.

The foregoing derivation and discussion of the modified interaction formulas should be supplemented by study of the foreword to the column load tables of the AISCM. In using the modified formulas a quick trial selection may be made on the basis of the following crude approximation for the equivalent load:

$$P_{eq} = P + B_x M_x + B_y M_y \tag{5.6}$$

Average values from the tables for B_x and B_y may be used for the initial trial selection. The use of Eq. (5.6) will result in an overestimate of the actual requirements; hence the initial trial should be taken so as to provide slightly less than the indicated value of P_{eq} by Eq. (5.6).

5.5 FLOWCHARTS

Flowchart 5.1

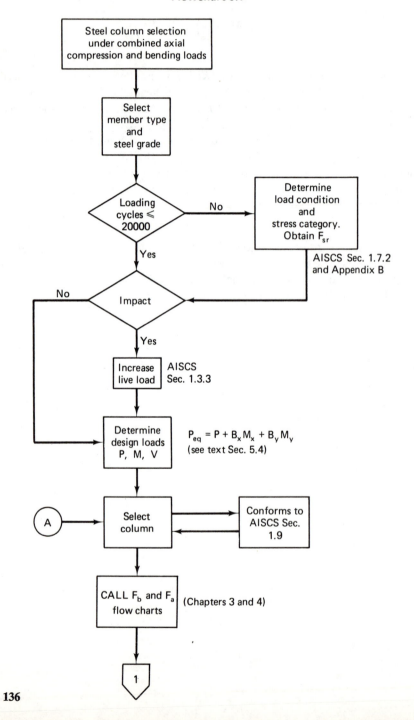

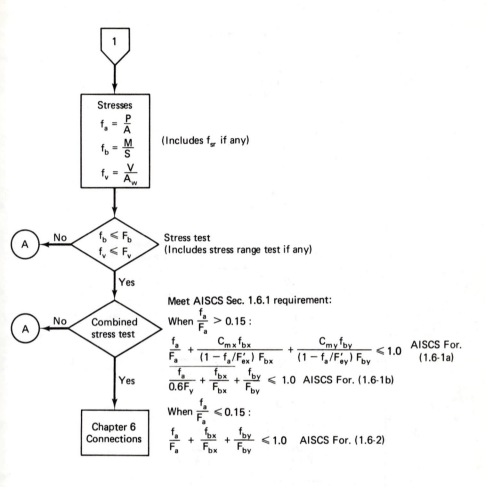

Stresses

$$f_a = \frac{P}{A}$$

$$f_b = \frac{M}{S}$$

$$f_v = \frac{V}{A_w}$$

(Includes f_{sr} if any)

$f_b \leqslant F_b$
$f_v \leqslant F_v$

Stress test
(Includes stress range test if any)

Combined stress test

Meet AISCS Sec. 1.6.1 requirement:

When $\dfrac{f_a}{F_a} > 0.15$:

$$\frac{f_a}{F_a} + \frac{C_{mx} f_{bx}}{(1 - f_a/F'_{ex}) F_{bx}} + \frac{C_{my} f_{by}}{(1 - f_a/F'_{ey}) F_{by}} \leqslant 1.0 \quad \begin{array}{l}\text{AISCS For.}\\ (1.6\text{-}1a)\end{array}$$

$$\frac{f_a}{0.6F_y} + \frac{f_{bx}}{F_{bx}} + \frac{f_{by}}{F_{by}} \leqslant 1.0 \quad \text{AISCS For. } (1.6\text{-}1b)$$

When $\dfrac{f_a}{F_a} \leqslant 0.15$:

$$\frac{f_a}{F_a} + \frac{f_{bx}}{F_{bx}} + \frac{f_{by}}{F_{by}} \leqslant 1.0 \quad \text{AISCS For. } (1.6\text{-}2)$$

Chapter 6 Connections

Flowchart 5.2

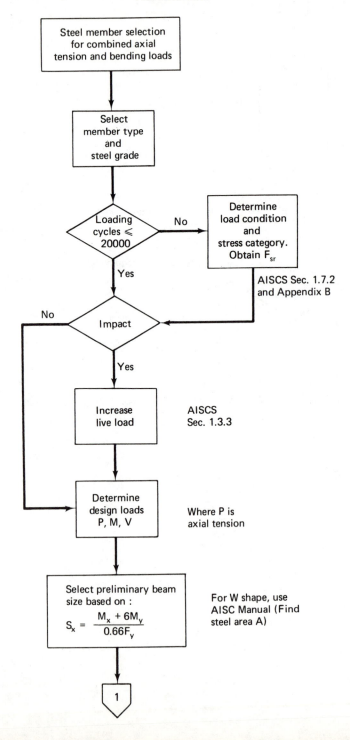

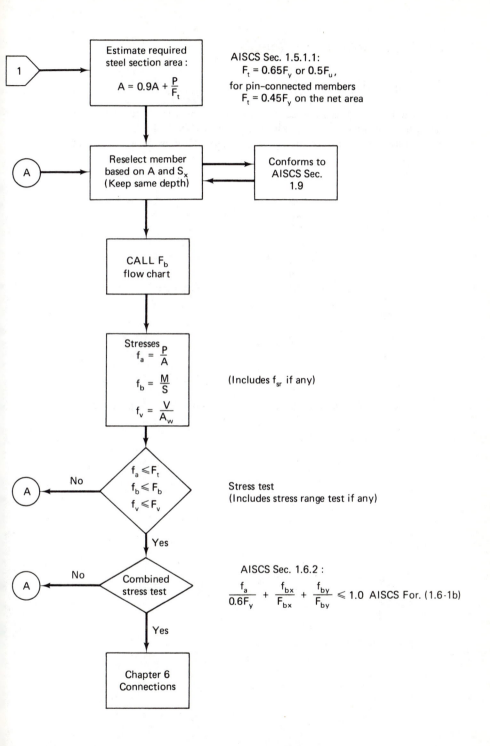

5.6 ILLUSTRATIVE EXAMPLES

Example 5.1†

A 30-ft-long column is subjected to an axial compression load, $P = 600$ kips, and lateral uniform load, $w = 0.30$ kip/ft, that causes bending about the column weak axis as shown. Select an economical W shape to satisfy the AISCS. Use A36 steel.

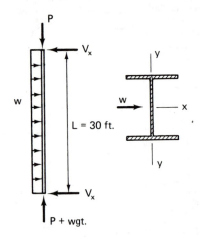

†An alternative solution of this problem will be found in Ref. 1.7, page 188.

Solution

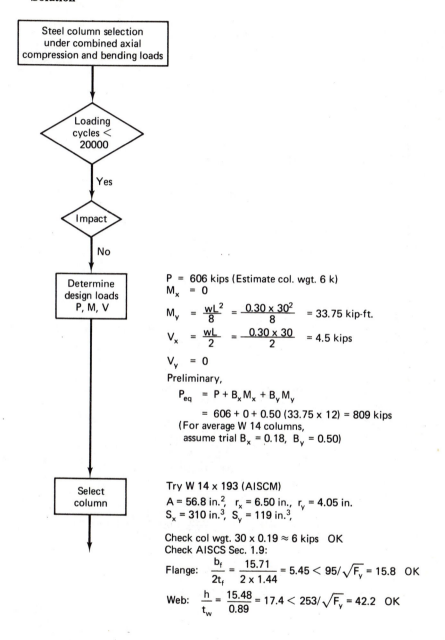

Steel column selection under combined axial compression and bending loads

Loading cycles < 20000

Yes

Impact

No

Determine design loads P, M, V

$P = 606$ kips (Estimate col. wgt. 6 k)

$M_x = 0$

$M_y = \dfrac{wL^2}{8} = \dfrac{0.30 \times 30^2}{8} = 33.75$ kip-ft.

$V_x = \dfrac{wL}{2} = \dfrac{0.30 \times 30}{2} = 4.5$ kips

$V_y = 0$

Preliminary,

$P_{eq} = P + B_x M_x + B_y M_y$

$= 606 + 0 + 0.50\,(33.75 \times 12) = 809$ kips

(For average W 14 columns, assume trial $B_x = 0.18$, $B_y = 0.50$)

Select column

Try W 14 x 193 (AISCM)

$A = 56.8$ in.^{2}, $r_x = 6.50$ in., $r_y = 4.05$ in.

$S_x = 310$ in.^{3}, $S_y = 119$ in.^{3},

Check col wgt. $30 \times 0.19 \approx 6$ kips OK

Check AISCS Sec. 1.9:

Flange: $\dfrac{b_f}{2t_f} = \dfrac{15.71}{2 \times 1.44} = 5.45 < 95/\sqrt{F_y} = 15.8$ OK

Web: $\dfrac{h}{t_w} = \dfrac{15.48}{0.89} = 17.4 < 253/\sqrt{F_y} = 42.2$ OK

F_{by} = 27 ksi (Compact Section)

Since $\dfrac{b_f}{2t_f}$ = 5.45 < 65/$\sqrt{F_y}$ = 10.8   OK

F_a = 14.33 ksi
(From AISCS Table 3-36, Appendix A)

for $\dfrac{K\ell}{r} = \dfrac{1.0\,(30 \times 12)}{4.05}$ = 88.9

$f_a = \dfrac{P}{A} = \dfrac{606}{56.8}$ = 10.67 ksi

$f_{by} = \dfrac{M_y}{S_y} = \dfrac{33.75 \times 12}{119}$ = 3.40 ksi

$f_{bx} = f_{vy} = 0$

$f_{vx} = \dfrac{V_x}{1.33A_f} = \dfrac{4.5}{1.33\,(15.71 \times 1.44)}$ = 0.150 ksi

Stress test OK

$\dfrac{f_a}{F_a} = \dfrac{10.67}{14.33} = 0.745 > 0.15$

$\dfrac{f_a}{F_a} + \dfrac{C_{mx}f_{bx}}{(1 - f_a/F'_{ex})\,F_{bx}} + \dfrac{C_{my}f_{by}}{(1 - f_a/F'_{ey})\,F_{by}}$

$= 0.745 + 0 + \dfrac{1.0 \times 3.40}{(1 - 10.67/18.89)\,27} = 1.034 > 1.0$ (not OK see note below)

(F'_{ey} = 18.89 ksi obtain from AISCS Table 9,
Appendix A for $K\ell_b/r_b$ = 88.9)

$\dfrac{f_a}{0.6F_y} + \dfrac{f_{bx}}{F_{bx}} + \dfrac{f_{by}}{F_{by}} = \dfrac{10.67}{0.6 \times 36} + 0 + \dfrac{3.40}{27} = 0.62 < 1.0$ OK

Note: Although the trial choice of a W 14 × 193 fails to meet the specification requirement, the overrun is only 3.4%. The practice in some situations may permit an overrun of a few percent. Alternatively, the next largest shape may be called for—a W 14 × 211—without repeating the calculations, since it has an area 9.3% greater than the trial selection. The same example appeared in the first edition of this text and a W 14 × 202 shape was found to be satisfactory. As of September 1, 1978, the rolling of the W 14 × 202 size was discontinued as part of a planned program of reduction and simplification of available sizes in mill practice.

Example 5.2

Same as Example 5.1, except the lateral uniform load causes bending about the column strong axis.

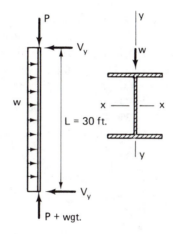

Solution

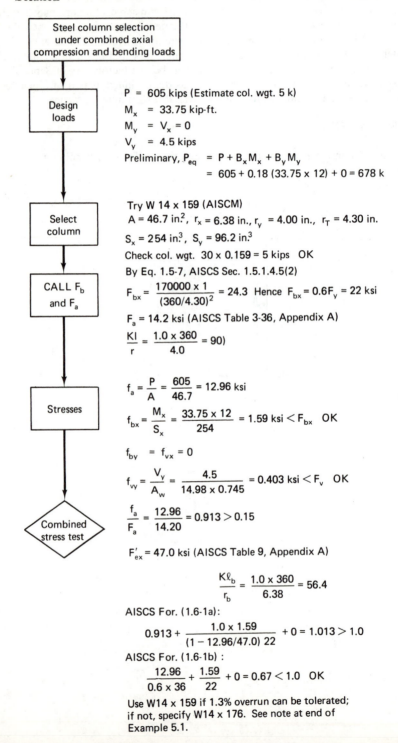

Steel column selection under combined axial compression and bending loads

Design loads

$P = 605$ kips (Estimate col. wgt. 5 k)
$M_x = 33.75$ kip-ft.
$M_y = V_x = 0$
$V_y = 4.5$ kips
Preliminary, $P_{eq} = P + B_x M_x + B_y M_y$
$\quad\quad\quad\quad = 605 + 0.18\,(33.75 \times 12) + 0 = 678$ k

Select column

Try W 14 x 159 (AISCM)
$A = 46.7$ in.^{2}, $r_x = 6.38$ in., $r_y = 4.00$ in., $r_T = 4.30$ in.
$S_x = 254$ in.^{3}, $S_y = 96.2$ in.^{3}
Check col. wgt. $30 \times 0.159 = 5$ kips   OK
By Eq. 1.5-7, AISCS Sec. 1.5.1.4.5(2)

CALL F_b and F_a

$F_{bx} = \dfrac{170000 \times 1}{(360/4.30)^2} = 24.3$ Hence $F_{bx} = 0.6F_y = 22$ ksi

$F_a = 14.2$ ksi (AISCS Table 3-36, Appendix A)
$\dfrac{Kl}{r} = \dfrac{1.0 \times 360}{4.0} = 90$)

Stresses

$f_a = \dfrac{P}{A} = \dfrac{605}{46.7} = 12.96$ ksi

$f_{bx} = \dfrac{M_x}{S_x} = \dfrac{33.75 \times 12}{254} = 1.59$ ksi $< F_{bx}$ OK

$f_{by} = f_{vx} = 0$

$f_{vy} = \dfrac{V_y}{A_w} = \dfrac{4.5}{14.98 \times 0.745} = 0.403$ ksi $< F_v$ OK

Combined stress test

$\dfrac{f_a}{F_a} = \dfrac{12.96}{14.20} = 0.913 > 0.15$

$F'_{ex} = 47.0$ ksi (AISCS Table 9, Appendix A)

$$\dfrac{K\ell_b}{r_b} = \dfrac{1.0 \times 360}{6.38} = 56.4$$

AISCS For. (1.6-1a):
$$0.913 + \dfrac{1.0 \times 1.59}{(1 - 12.96/47.0)\,22} + 0 = 1.013 > 1.0$$

AISCS For. (1.6-1b) :
$$\dfrac{12.96}{0.6 \times 36} + \dfrac{1.59}{22} + 0 = 0.67 < 1.0 \quad \text{OK}$$

Use W14 x 159 if 1.3% overrun can be tolerated; if not, specify W14 x 176. See note at end of Example 5.1.

Example 5.3

Select an economical W shape for the same load conditions as in Example 5.2 but with adequate lateral supports provided along the column to cause the member to deflect in the plane of its minor axis.

Solution

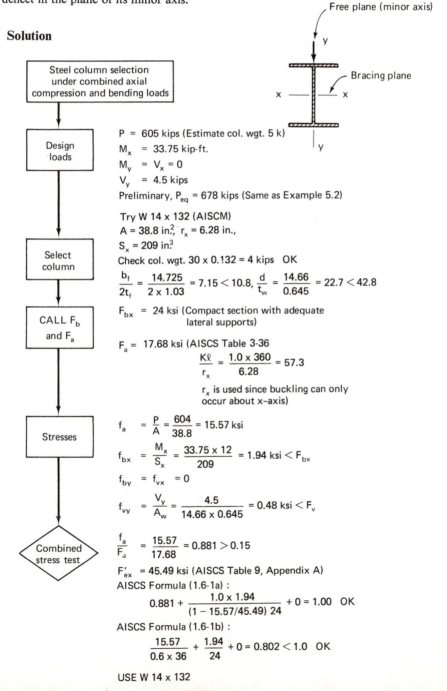

Steel column selection under combined axial compression and bending loads

Design loads

P = 605 kips (Estimate col. wgt. 5 k)
M_x = 33.75 kip-ft.
$M_y = V_x = 0$
V_y = 4.5 kips
Preliminary, P_{eq} = 678 kips (Same as Example 5.2)

Select column

Try W 14 x 132 (AISCM)
A = 38.8 in.^{2}, r_x = 6.28 in.,
S_x = 209 in.^{3}
Check col. wgt. 30 x 0.132 = 4 kips   OK
$\frac{b_f}{2t_f} = \frac{14.725}{2 \times 1.03} = 7.15 < 10.8$, $\frac{d}{t_w} = \frac{14.66}{0.645} = 22.7 < 42.8$

CALL F_b and F_a

F_{bx} = 24 ksi (Compact section with adequate lateral supports)

F_a = 17.68 ksi (AISCS Table 3-36
$$\frac{K\ell}{r_x} = \frac{1.0 \times 360}{6.28} = 57.3$$

r_x is used since buckling can only occur about x–axis)

Stresses

$f_a = \frac{P}{A} = \frac{604}{38.8} = 15.57$ ksi

$f_{bx} = \frac{M_x}{S_x} = \frac{33.75 \times 12}{209} = 1.94$ ksi $< F_{bx}$

$f_{by} = f_{vx} = 0$

$f_{vy} = \frac{V_y}{A_w} = \frac{4.5}{14.66 \times 0.645} = 0.48$ ksi $< F_v$

Combined stress test

$\frac{f_a}{F_a} = \frac{15.57}{17.68} = 0.881 > 0.15$

F'_{ex} = 45.49 ksi (AISCS Table 9, Appendix A)
AISCS Formula (1.6-1a) :
$$0.881 + \frac{1.0 \times 1.94}{(1 - 15.57/45.49) \, 24} + 0 = 1.00 \quad OK$$

AISCS Formula (1.6-1b) :
$$\frac{15.57}{0.6 \times 36} + \frac{1.94}{24} + 0 = 0.802 < 1.0 \quad OK$$

USE W 14 x 132

Example 5.4

Select a column in a building frame for a 24-ft story height to support a 260-kip axial compression load and 100 and 120 kip-ft bending moments acting at the top and bottom ends, respectively, as shown. The column is braced against sidesway in the weak bending plane of the column (*xz* plane) but with possible sidesway in the strong bending plane (*yz* plane). Use AISCS and A36 steel.

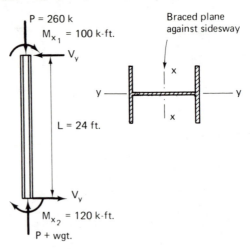

Solution

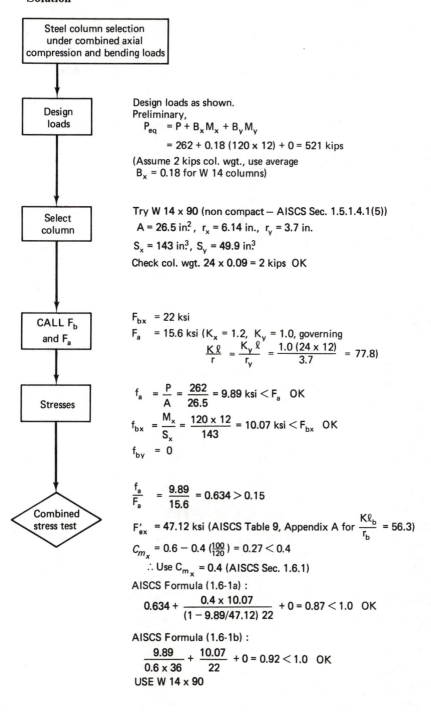

Steel column selection under combined axial compression and bending loads

Design loads

Design loads as shown.
Preliminary,

$$P_{eq} = P + B_x M_x + B_y M_y$$

$$= 262 + 0.18 (120 \times 12) + 0 = 521 \text{ kips}$$

(Assume 2 kips col. wgt., use average
$B_x = 0.18$ for W 14 columns)

Select column

Try W 14 x 90 (non compact — AISCS Sec. 1.5.1.4.1(5))
$A = 26.5 \text{ in}^2$, $r_x = 6.14 \text{ in.}$, $r_y = 3.7 \text{ in.}$
$S_x = 143 \text{ in}^3$, $S_y = 49.9 \text{ in}^3$
Check col. wgt. 24 x 0.09 = 2 kips OK

CALL F_b and F_a

$F_{bx} = 22 \text{ ksi}$
$F_a = 15.6 \text{ ksi}$ ($K_x = 1.2$, $K_y = 1.0$, governing

$$\frac{K\ell}{r} = \frac{K_y \ell}{r_y} = \frac{1.0 (24 \times 12)}{3.7} = 77.8)$$

Stresses

$$f_a = \frac{P}{A} = \frac{262}{26.5} = 9.89 \text{ ksi} < F_a \quad \text{OK}$$

$$f_{bx} = \frac{M_x}{S_x} = \frac{120 \times 12}{143} = 10.07 \text{ ksi} < F_{bx} \quad \text{OK}$$

$$f_{by} = 0$$

Combined stress test

$$\frac{f_a}{F_a} = \frac{9.89}{15.6} = 0.634 > 0.15$$

$F'_{ex} = 47.12 \text{ ksi}$ (AISCS Table 9, Appendix A for $\dfrac{K\ell_b}{r_b} = 56.3$)

$C_{m_x} = 0.6 - 0.4 \left(\frac{100}{120}\right) = 0.27 < 0.4$

$\therefore$ Use $C_{m_x} = 0.4$ (AISCS Sec. 1.6.1)

AISCS Formula (1.6-1a) :

$$0.634 + \frac{0.4 \times 10.07}{(1 - 9.89/47.12) \, 22} + 0 = 0.87 < 1.0 \quad \text{OK}$$

AISCS Formula (1.6-1b) :

$$\frac{9.89}{0.6 \times 36} + \frac{10.07}{22} + 0 = 0.92 < 1.0 \quad \text{OK}$$

USE W 14 x 90

Example 5.5

A simply supported member has a span of 20 ft and carries a uniform live and dead load of 2.56 kips/ft (including its own weight), and axial *tension* load of 65 kips acting through the centroid of the member as shown. The compression flange of the member is laterally supported against local buckling. Select an economical W shape to satisfy the AISCS. Use A36 steel.

Solution

Steel member selection under combined axial tension and bending loads

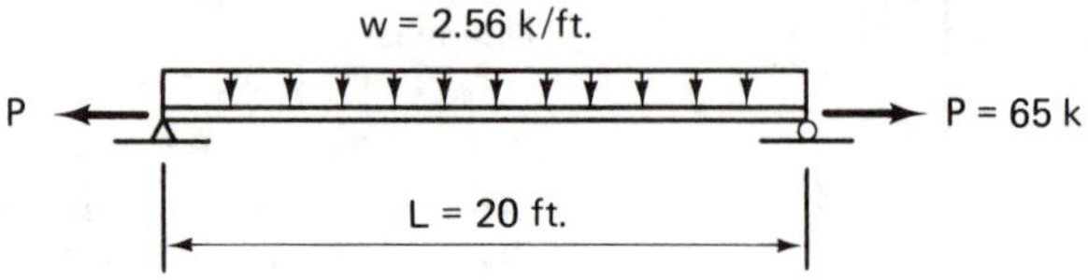
w = 2.56 k/ft.
P
P = 65 k
L = 20 ft.

Design loads

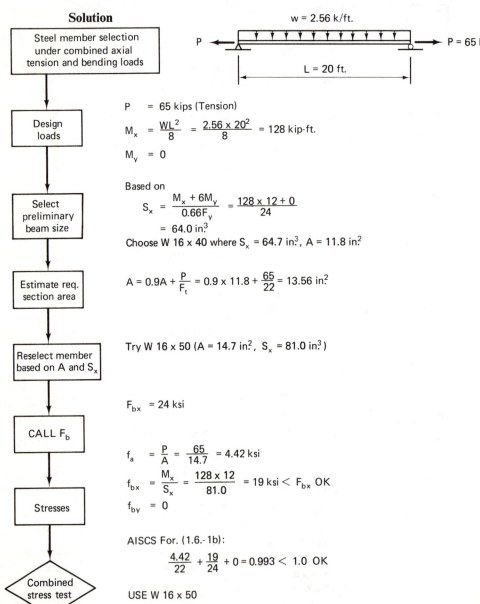

$P = 65$ kips (Tension)

$$M_x = \frac{WL^2}{8} = \frac{2.56 \times 20^2}{8} = 128 \text{ kip-ft.}$$

$M_y = 0$

Select preliminary beam size

Based on

$$S_x = \frac{M_x + 6M_y}{0.66F_y} = \frac{128 \times 12 + 0}{24} = 64.0 \text{ in.}^3$$

Choose W 16 x 40 where $S_x = 64.7$ in.^{3}, $A = 11.8$ in.^{2}

Estimate req. section area

$$A = 0.9A + \frac{P}{F_t} = 0.9 \times 11.8 + \frac{65}{22} = 13.56 \text{ in.}^2$$

Reselect member based on A and S_x

Try W 16 x 50 ($A = 14.7$ in.^{2}, $S_x = 81.0$ in.^{3})

CALL F_b

$F_{bx} = 24$ ksi

Stresses

$$f_a = \frac{P}{A} = \frac{65}{14.7} = 4.42 \text{ ksi}$$

$$f_{bx} = \frac{M_x}{S_x} = \frac{128 \times 12}{81.0} = 19 \text{ ksi} < F_{bx} \text{ OK}$$

$f_{by} = 0$

Combined stress test

AISCS For. (1.6.-1b):

$$\frac{4.42}{22} + \frac{19}{24} + 0 = 0.993 < 1.0 \text{ OK}$$

USE W 16 x 50

PROBLEMS

5.1, 5.2, 5.3, 5.4, and 5.5. Rework Examples 5.1 through 5.5, changing the yield point of the steel from 36 to 50 ksi.

5.6. On the basis of allowable stress, ignoring column behavior and the effect of deflection on moment, select a stub cantilever for the conditions illustrated in Fig. 5.1, if the length is 40 in., load P is 400 kips, with an eccentricity $e = 6$ in., using a W shape of A36 steel bent about the strong axis. Check the adequacy of the selection by use of the AISCS interaction equation.

5.7. Redesign for the situation described in Problem 4.6, but with a lateral load of 250 lb/ft over the 32-ft column length.

5.8. A square box column, 38 ft in length, with sidewalls 26 in. apart (dimensioned as shown to the middle planes of the plates), carries an axial load of 600 kips and a lateral load of 300 lb/ft, acting normal to one of the flat sides. Determine the plate thickness to the nearest $\frac{1}{16}$ in. if A36 steel is used.

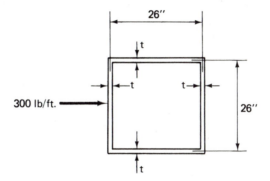

5.9. Redesign for the conditions of Problem 3.14 if the beam, in addition to the specified vertical and horizontal beam loads, carries a compressive force of 200 kips applied through the centroid at each end.

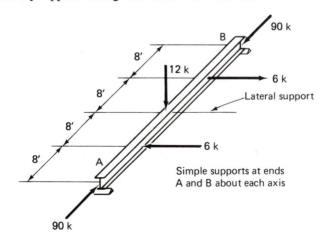

5.10. The beam in the figure is restrained laterally in the weak direction by ties as shown. For bending in the weak (horizontal) directions, it may be assumed as hinged at the center. Select an adequate member of A36 steel for this biaxial bending situation.

6

CONNECTIONS

6.1 INTRODUCTION

The component rolled plates and shapes of a steel structure are held together
by means of fasteners (rivets or bolts) or by welds, which may either fuse the
parts together into an integral unit or stitch them together intermittently as
do fasteners. The fasteners and welds are used in the shop fabrication processes to build up members and again in the field erection to connect the
separate members together to form the completed structural frame. If a
member is too large to be shipped as a single unit, field connections must
also be used to splice the member segments. Because of the relatively greater
cost of making field connections, such splices are kept to a minimum.

In truss construction, the tension and compression members meeting at
a joint may be attached separately by fasteners to a *gusset plate* (see Fig. 2.3),
or, if welding is used, it may be possible to join them directly without the use
of an auxiliary plate.

In building-frame construction the AISCS, Sec. 1.2, recognizes three
basic types of beam-to-column framing.

Type 1, commonly designated as *rigid frame* or *continuous frame*, has
beam-to-column or beam-to-girder connections that transmit the calculated
moments and shears and have sufficient rigidity to provide the full continuity that has been assumed in the analysis. This means that no local kinks
will be formed at allowable loads due to premature local yielding.

Type 2, designated as *simple framing*, provides flexible connections that
have adequate shear strength without developing appreciable moment. The
connections are purposely designed so as to permit beam end rotations
relative to the column or girder to a degree that will permit one to ignore the

incidental bending moments and small inelastic yielding that may be developed.

Type 3, *semirigid framing*, "assumes that the connections . . . possess a dependable and known moment capacity intermediate in degree between the rigidity of Type 1 and the flexibility of Type 2."

Type 1 moment-resisting connections are frequently used in the main building frames to resist wind and earthquake forces, with Type 2 connections being used in the remainder of the structure.

In addition to the material on connections in the AISCS and Commentary, the AISCM provides design information and tables that facilitate the proportioning of commonly used connections.

A connection is said to be loaded concentrically if the resultant (axial) force passes through the centroid of the fastener group or weld pattern. At low loads the distribution of stress transfer in such a connection is quite nonuniform, but prior to failure the local yielding of material tends to distribute the load uniformly to all parts of the connection. This fact, together with the results of experience and many laboratory tests to failure, permits the assumption in design that all parts of a concentrically loaded connection share equally in resisting the applied force. If the resultant force does not pass through the centroid of the connection, the force system may be reduced to a force and a couple at the connection centroid. Each element of the connection is assumed to resist the axial component of force uniformly and to resist the moment in proportion to the distance of that element from the centroid of the connection. Design rules based on the foregoing assumptions have been found to be safe and workable.

Design recommendations and allowable stresses in connections as supplied in the AISCS are very largely backed up and based on recommendations of the Research Council on Riveted and Bolted Joints and the American Welding Society, for bolted and welded connections, respectively.

Four methods of making structural connections (rivets, bolts, pins, and welds) will be discussed and simple design examples presented. The information available on this subject in the AISCM also should be studied in detail.

The overall strength and safety of a structure may be directly dependent on the connections that join main members. Such connections should be shown explicitly and in detail on the design drawings, both in the interest of safe design and for economy in view of bid-price allowances that might otherwise be made for contingencies.

6.2 RIVETED AND BOLTED CONNECTIONS

For many years riveting was the accepted method for making connections. However, the use of rivets has declined rapidly due to the development and economic advantages of welding and high-strength bolts. The advent of both welding and high-strength bolting has made possible the advantageous use of a combination of these fastening methods, such as shop fabrication by welding followed by the use of high-strength bolts for the field connections. In this way the advantages of each procedure are realized, as the welding is under shop-controlled conditions with the members positioned to produce good welds and economy in fabrication. The advantages of rapid assembly while the members are being held in position in the field are obtained through the use of high-strength bolts.

Rivets and bolts, as shown in Fig. 6.1, transmit force from one plate element to another by either single shear or double shear. At low loads the transfer of force from plate to plate is largely by friction. At higher loads, after slip takes place, the fasteners come into direct bearing. In more complex joints with multiple plates interacting, more than two shear planes may occur.

Fig. 6.1 Rivets and bolts.

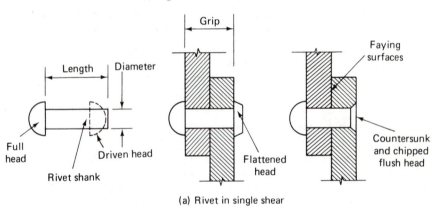

(a) Rivet in single shear

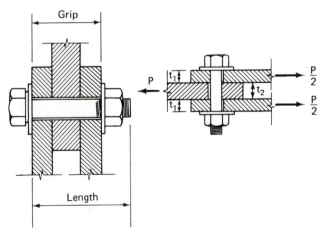

(b) Bolt in double shear

Rivets are manufactured with a special full (acorn) head and are installed in holes that are punched or drilled $\frac{1}{16}$ in. larger in diameter than the rivets. Hot driven rivets are brought to approximately 1800°F before being placed in the hole, and a second head is formed by means of a riveting hammer or a machine pressure type of riveter. During the process of forming this head, the shank is upset so that it should now completely fill the hole (see Fig. 6.1). The formed head can be either a full button head, a flattened head, or a countersunk and chipped head. The chipping is usually done with a chisel and air hammer after the rivet has cooled. Rivets with countersunk and chipped heads have less strength and cost considerably more than those with full or flattened heads. During cooling, the rivet shrinks, setting up tensile forces in the shank which may approach the yield point of the material. This residual tension force is unpredictable and may vary from practically nothing to a stress equal to the yield point. Because of this uncertainty, the clamping force exerted on the joined material is neglected in design calculations.

Rivets shall conform to the provisions of the "Specifications for Structural Rivets," ASTM A502, grades 1 or 2. The size of rivets used in ordinary steel construction ranges from $\frac{5}{8}$ to $1\frac{1}{2}$ in. in diameter by $\frac{1}{8}$-in. increments. The allowable stresses are tabulated in AISCS Table 1.5.2.1. The AISCM should now be studied with regard to detailing practice, erection clearances, recommended spacing, and conventional signs for use on drawings for both rivets and bolts.

Unfinished bolts, also known as ordinary or common bolts, shall conform to the "Specifications for Low Carbon Steel Externally and Internally Threaded Standard Fasteners," ASTM A307. These bolts range from $\frac{5}{8}$ to $1\frac{1}{2}$ in. in diameter by $\frac{1}{8}$-in. increments. The allowable stresses are also tabulated in the AISCS Table 1.5.2.1. Because of the uncertainty as to whether or not the threaded portion of unfinished bolts extends into the shear plane, their allowable stresses are considerably less than those permitted for rivets or high-strength bolts. Their use is usually restricted to structures subjected to static loads, and for secondary members such as purlins, girts, and bracing. AISCS, Sec. 1.15.12, lists specific connection types for which A307 bolts may *not* be used.

Since the initial tension developed by rivets and unfinished bolts is uncertain and possibly very small, no frictional resistance on the faying surfaces is assumed, and slip may occur at low shearing loads. This brings the rivets or bolts into bearing, and the mode of stress transfer is as shown in Fig. 6.2.

The bearing stress f_p on the contact area between the fasteners and the connected plates is defined as the transmitted shear load P divided by the total effective bearing area, dt, where d is the nominal diameter of the rivet or unfinished bolt and t the thickness of the connected material.

The shearing stress f_v in fasteners is defined as the transmitted shear load

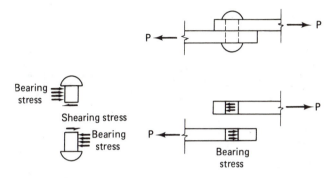

Fig. 6.2 Stress transfer by shear and bearing in a riveted or bolted bearing-type connection.

P divided by the total effective shearing area A_v:

$$f_v = \frac{P}{A_v}$$ SHEARING STRESS

where

$$A_v = \begin{cases} \dfrac{\pi d^2}{4} & \text{for single shear} \\[2ex] \dfrac{\pi d^2}{2} & \text{for double shear} \end{cases}$$

Rivets and unfinished bolts are acceptable in tension-type connections (such as hangers) for static loads and the allowable stresses are tabulated in AISCS, Table 1.5.2.1. Stresses for rivets are based on the gross cross-sectional area using the nominal diameter, and for bolts on the area determined by the maximum diameter of the threaded portion.

High-strength bolts are torqued to a high tensile stress in the shank, thus developing a dependable clamping pressure. Shearing stress is transferred by friction at working loads, as illustrated in Fig. 6.3. High-strength bolts are a preferred fastener for field connections, resistance to stress reversal, impact loads, and other applications where joint slip is not desirable. Ease of installation is another desirable attribute.

High-strength bolts are available in two different strength levels, and are used in accordance with the provisions of the "Specification for Structural Joints Using ASTM A325 or A490 Bolts" as approved by the Research Council on Riveted and Bolted Structural Joints. A449 bolts have the same chemistry as A325 bolts but are used only in bearing-type high-strength joints requiring diameters greater than $1\frac{1}{2}$ in. or as anchor bolts or threaded rods.

In the friction-type connection, the bolts are not actually stressed in shear and are not in bearing, since no slip occurs at allowable loads. However, a

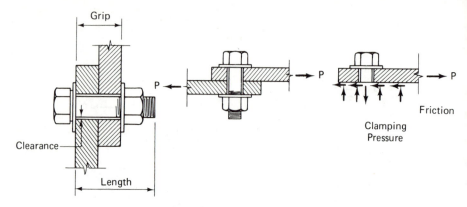

Fig. 6.3 Friction-type connection.

shear stress is specified as a matter of convenience, and the number of fasteners is determined in the same manner as for other riveted or bolted connections. High-strength bolts are tightened so as to produce a minimum initial tension in the bolt shank equal to the proof load, or approximately 70% of the tensile strength of the bolt. In order to obtain the specified initial tension in the bolts, they are usually tightened with calibrated wrenches or by the turn-of-nut method. A third method of installation was added to the Bolt Specification in 1972, whereby a direct tension indicator is utilized. Specifications require that hardened washers be used under the turned element when using the calibrated-wrench method and for A490 bolts when using the turn-of-nut method. Hardened washers are required under both head and nut when using A490 bolts to connect material with a yield point of less than 40 ksi. Beveled washers are required when an outer face of the connection has a slope greater than 1: 20. In the turn-of-nut method, nuts are first brought to a "snug" fit, defined as the condition when all surfaces are in good contact and no free rotation of the nut is possible. Bolts under eight diameters or 8 in. in length are then given an additional one-half turn of the nut; over eight diameters or 8 in., two-thirds of a turn; and if both faces have a 1: 20 slope, without use of washers, the nuts are to be given a three-quarter turn, regardless of length.

Resistance to slip is determined by the amount of bolt tension and the condition of the contact surfaces in a given connection. Connections having painted contact surfaces or contact surfaces of unrusted mill scale offer the least resistance to slip; rusted surfaces that have been well cleaned may provide up to two times as much resistance.

Rivets and bolts in combined shear and tension are proportioned according to AISCS, Sec. 1.6.3. In the case of rivets and bearing-type bolted connections, the allowable tensile stresses are reduced if the simultaneous shear stress exceeds a certain value. For example, the allowable tensile stress for A502 grade 1 rivets is

$$F_t = 30.0 - 1.3f_v \leq 23.0 \text{ ksi}$$

This says, in effect, that the allowable tensile stress is in no case greater than 23 ksi, and if the <u>shear stress</u> exceeds 5.4 ksi, allowable tensile stress will be less than 23 ksi, as given by the formula. Similar formulas for grade 2 rivets, and for unfinished and high-tensile bolts in bearing-type joints, are provided in Sec. 1.6.3 of the AISCS.

In the case of friction-type joints with high-tensile bolts, the allowable shear stress is dependent on the existence of adequate initial tension. Hence, if additional tension due to external loads is applied, the clamping force is reduced and AISCS, Sec. 1.6.3, requires that "the maximum shear stress allowed by Table 1.5.2.1 (AISCS) shall be multiplied by the reduction factor $(1 - f_t A_b/T_b)$, where f_t is the average tensile stress due to a direct load applied to all of the bolts in a connection and T_b is the specified pretension load of the bolt" (see AISCS, Table 1.23.5).

At this point the reader should review in detail the AISCS and AISCM information on rivets and bolts, as follows: AISCS, Secs. 1.5.2.1 and 1.5.2.2, together with Table 1.5.2.1. Also read Sec. 1.16 of AISCS, which provides detailed information on minimum pitch, minimum edge distance, and so forth, and Secs. 1.23.4 and 1.23.5, which define good fabrication practice for riveted and bolted joints. Finally, note the detailed dimension and weight information regarding rivets and bolts in the AISCM, and the tabulation of rivet and bolt strength values, at prescribed allowable stresses. A reason for the textual omission of much of the information just referred to is the desirability of fostering the direct use of specification and manual information in the handling of design problems.

Examples 6.1, 6.2, and 6.3 now illustrate the design of simple concentrically loaded riveted and bolted joints.

Example 6.1 *Lap Joints*

A lap joint is simply a joint in which two members overlap and are connected to each other with some type of fastener. The lap joint is not a desirable structural connection and should be used only for minor connections. The eccentricity of the loads causes secondary bending stresses in the members and at least two fasteners should be used on each line. (Edge distance, net section, and spacing checks omitted for these; see Example 6.2B.)

A. Two $\frac{1}{4} \times 8$-in. A36 plates are to be connected using $\frac{3}{4}$-in. A502 grade 1 rivets, as shown in the figure. How many rivets are required to transfer the load $P = 35$ kips?

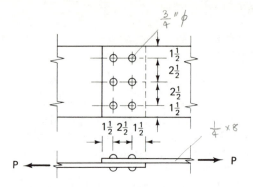

Solution

Rivets are critical in single shear or bearing (refer to AISCS, Secs. 1.5.1.5.3, 1.5.2.1):

$$\text{bearing value} = 1.5 \times 58 \times \tfrac{3}{4} \times \tfrac{1}{4} = 16.3 \text{ kips/rivet} \qquad (F_p = 1.5F_u)$$

$$\text{shear value} = 17.5 \times 0.442 = 7.74 \text{ kips/rivet} \qquad (F_v = 17.5 \text{ ksi})$$

Single shear governs:

$$\text{number of rivets req'd} = \frac{35}{7.74} = 4.5$$

Use six rivets for symmetry. (Symmetry in the placement of fasteners is desirable in order to avoid secondary stresses that add to the already complex stress distribution. Note that rivet values could have been obtained directly from the AISCM.)

B. Same as (A), except use $\tfrac{3}{4}$-in.-diameter A325 bolts in a friction-type connection with standard-size holes.

Solution

$$\text{shear value} = F_v A_b = 17.5 \times 0.442 = 7.74 \text{ kips/bolt}$$

$$\text{number of bolts req'd} = \frac{35}{7.74} = 4.5$$

Use six bolts, same pattern as for Example 6.1A.

C. Same as (B), except in bearing-type connection with threads excluded from shear plane (see AISCS, Secs. 1.5.1.5.3, 1.5.2.1).

Solution

$$\text{bearing on plate} = 1.5 \times 58 \times \tfrac{3}{4} \times \tfrac{1}{4} = 16.3 \text{ kips/bolt}$$

$$\text{shear value} = 30 \times 0.442 = 13.26 \text{ kips/bolt} \qquad (\text{governs})$$

$$\text{number of bolts req'd} = \frac{35}{13.26} = 2.6$$

Use four bolts, in two rows (4 in. apart) and two lines.

Example 6.2 *Butt Splice*

A. A butt splice is to be designed as shown in the figure. Use $\frac{3}{4}$-in.-diameter A325 bolts with standard-size holes in a friction-type connection and A36 steel. $P = 90$ kips. Determine the number of bolts required on each side of the splice. Bolts are in double shear. (Spacing and net section checks omitted; see Example 6.2B.)

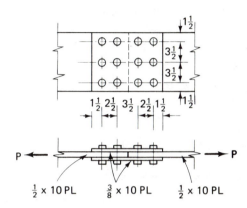

Solution

capacity of one bolt in double shear $= 2F_vA_b = 2 \times 17.5 \times 0.442 = 15.47$ kips

bearing capacity $= 1.5 \times 58 \times 0.5 \times 0.75 = 32.6$ kips

$$\text{number of bolts req'd} = \frac{90}{15.47} = 5.82$$

Use six bolts (three bolts per line).

B. Same as (A), except use $\frac{3}{4}$-in.-diameter A325 bolts in a bearing-type connection with threads excluded from the shear plane. Increase P to 130 kips.

Solution

capacity of one bolt in double shear $= 2 \times 0.442 \times 30 = 26.52$ kips

(governs)

capacity of one bolt in bearing on $\frac{1}{2}$-in. plate $= 1.5 \times 58 \times 0.75 \times \frac{1}{2} = 32.6$ kips

$$\text{number of bolts req'd} = \frac{130}{26.52} = 4.90$$

Use six bolts, as in part A.

Check the tensile capacity (AISCS Sec. 1.5.1.1):

Net section capacity ($F_t = 0.5F_u = 29$ ksi):

$$[10 - 3(\tfrac{3}{4} + \tfrac{1}{8})] \times \tfrac{1}{2} \times 29 = 106.9 \text{ kips} < 130$$

Gross section capacity ($F_t = 0.6F_y = 22$ ksi):

$$10 \times \tfrac{1}{2} \times 22 = 110 \text{ kips} \qquad \text{(net section governs)}$$

Change to $\tfrac{5}{8}$ plate.

By proportion, the net section capacity is

$$\frac{\tfrac{5}{8}}{\tfrac{1}{2}} \times 106.9 = 133.6 \text{ kips} > 130 \qquad \text{OK}$$

Check the edge distance (see AISCS, Sec. 1.16.5.2):

$$\frac{2P}{F_u t} = \frac{2 \times 130}{6 \times 58 \times \tfrac{5}{8}} = 1.20 < 1.5 \qquad \text{OK}$$

The value 1.5 also satisfies Table 1.16.5.1.

Minimum spacing: (AISCS, Sec. 1.16.4.1):

$$\frac{2P}{F_u t} + \frac{d}{2} = 1.20 + \frac{3}{4 \times 2} = 1.58 < 2.5 \qquad \text{OK}$$

or

$$2.67 \times \tfrac{3}{4} = 2.00 < 2.5 \qquad \text{OK}$$

Example 6.3 *Bracket Connection*

A tension member, two L $4 \times 3 \times \tfrac{1}{2}$, carries a tension load P of 120 kips with a direction of 30° from the horizontal axis. A bracket utilizing a structural tee section will be used to connect the tension member by rivets and will be joined to a column flange by high-strength bolts with a friction-type connection, as shown in the figure. Determine the required size and number of rivets and bolts. Use A502 grade 2 rivets, A490 high-strength bolts, and A36 steel in accordance with the AISCS.

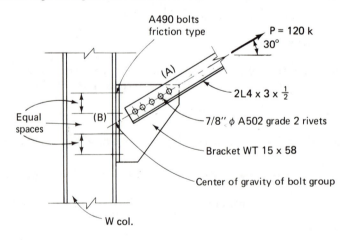

Solution

A. *Connection A:* tension angles to bracket (try WT 15×58, $t_w = 0.565$ in.). The allowable shear stress is

$$F_v = 22 \text{ ksi} \qquad \text{(A502 grade 2, AISCS, Table 1.5.2.1)}$$

The allowable bearing stress is

$$F_p = 1.5 \times 58 = 87 \text{ ksi} \qquad \text{(AISCS, Sec. 1.5.1.5.3)}$$

Try $\frac{7}{8}$-in.-diameter rivets ($A_b = 0.601 \text{ in.}^2$).

The double shear strength is

$$2F_v A_b = 2 \times 22 \times 0.601 = 26.4 \text{ kips/rivet} \qquad \text{(governs)}$$

The bearing strength on the tee web equals

$$87 \times 0.565 \times \tfrac{7}{8} = 43.0 \text{ kips/rivet}$$

Check the net section capacity of the double-angle member:

$$\text{net area} = \underbrace{6.5}_{} - \overbrace{2 \times 0.5(\tfrac{7}{8} + \tfrac{1}{8})}^{\text{Area of 2 angles}} = 5.5 \text{ in.}^2.$$

$$\text{effective net area } A_e = 0.85 \times 5.5 = 4.68 \text{ in.}^2.$$

$$\text{(AISCS, Sec. 1.14.2.2)}$$

$$\text{tensile capacity} = 4.68 \times 29 = 135.7 \text{ kips} > 120 \qquad \text{OK}$$

$$\text{number of rivets req'd} = 120/26.4 = 4.55$$

Use five rivets. Arrange as shown, at $2\frac{1}{2}$ in. center to center. Detailer to check layout for clearances and edge distances.

B. *Connection B:* bracket to column flange. Assume that the tension load passes through the center of gravity of the bolts. Then the shear and tension components of the tension are:

Tension component:

$$T = P \cos 30° = 120 \times 0.866 = 104 \text{ kips}$$

Shear component:

$$V = P \sin 30° = 120 \times 0.5 = 60 \text{ kips}$$

The allowable stresses for A490 high-strength bolts in friction-type connection are:

Allowable tensile stress:

$$F_t = 54 \text{ ksi} \qquad \text{(AISCS, Table 1.5.2.1)}$$

Allowable shear stress:

$$F_v = 22 \text{ ksi} \qquad \text{(subject to reduction)}$$

Bolt tensile stress—assume eight bolts of $\frac{7}{8}$ in. diameter:

$$f_t = \frac{T}{nA_b} = \frac{104}{8 \times 0.601} = 21.6 \text{ ksi} < F_t \qquad \text{OK}$$

Bolt shear stress:

$$f_v = \frac{V}{nA_b} = \frac{60}{8 \times 0.601} = 12.5 \text{ ksi}$$

The pretension of a $\frac{7}{8}$-in. bolt, $T_b = 49$ kips (AISCS, Table 1.23.5). The reduced allowable shear stress due to combined tension is

$$F_v = 22\left(1 - \frac{f_t A_b}{T_b}\right) \quad \text{(see text and AISCS, Sec. 1.6.3)}$$

$$= 22\left(1 - 21.6 \times \frac{0.601}{49}\right) = 16.2 \text{ ksi} > f_v \quad \text{OK}$$

Use eight bolts of $\frac{7}{8}$ in. diameter.

Note that the prying force on the A490 bolts should also be investigated. The calculations involved are demonstrated in Example 6.13.

6.3 PINNED CONNECTIONS

Pinned connections are sometimes used in bridge-bearing supports with the purpose of permitting end rotation. They are also used to connect pin-connected members of the types discussed in Chapter 2. Ranging from 2 to 10 in. or more in diameter, they are designed in a manner analogous to bearing connections of bolts, but with lower allowable stresses and with the added requirement to check stress due to bending in the pin itself. Details of standard pins and caps or nuts to hold them in position are provided in the connection section of the AISCM. Allowable stresses due to shear, bending, and in bearing are $0.4F_y$, $0.75F_y$, and $0.9F_y$, respectively, as stated in AISCS, Secs. 1.5.1.2, 1.5.1.4.3, and 1.5.1.5.1, respectively. Although the actual distribution of stress in a short circular beam is complex, designs have been found to be satisfactory when based on simple beam theory and on average stress due to shear and bearing. Bending moments may be conservatively calculated on the assumption that forces are concentrated at the centers of bearing areas.

On the basis of assumed locations of acting forces, bending moments and shears may be determined, and a preliminary selection of required pin diameters may then be determined on the basis of whichever stress (bending or shear) is critical. The bearing stress may then be checked, and revision of either the pin diameter or length of bearing may then be made if required.

In bending,

$$f_b = \frac{M}{S} = \frac{32M}{\pi d^3}$$

Hence to maintain $f_b \leq F_b$, the required pin diameter is

$$d_{\text{req}} \geq \sqrt[3]{\frac{32M}{\pi F_b}} \tag{6.1}$$

In shear,

$$f_{v(\text{av})} = \frac{4V}{\pi d^2}$$

and, similarly, the required pin diameter for shear is

$$d_{\text{req}} \geq \sqrt{\frac{4V}{\pi F_v}} \tag{6.2}$$

Example 6.4 *Pin Connection*

Although the connection shown is of a type not commonly used in recent years, it illustrates the essential problems in pin design. For A36 steel the allowable stresses are, by reference to the AISCS sections referred to in the previous paragraph:

For shear:
$$F_v = 0.4F_y = 14.5 \text{ ksi}$$

for bending:
$$F_b = 0.75F_y = 27.0 \text{ ksi}$$

For bearing:
$$F_p = 0.90F_y = 33.0 \text{ ksi}$$

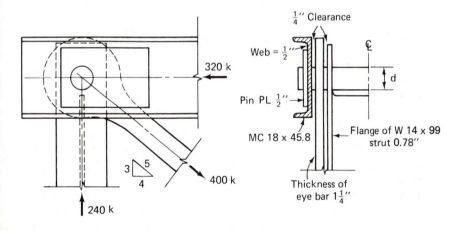

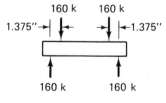

Horizontal components

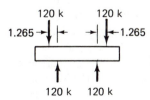

Vertical components

Bending moment:

$$M_h = 160 \times 1.375 = 220 \text{ kip-in.}$$

$$M_v = 120 \times 1.265 = 152 \text{ kip-in.}$$

$$M = \sqrt{220^2 + 152^2} = 267 \text{ kip-in.}$$

Pin diameter required by moment [Eq. (6.1)]:

$$d = \sqrt[3]{\frac{32M}{\pi F_b}} = \sqrt[3]{\frac{32 \times 267}{\pi \times 27}} = 4.65 \text{ in.}$$

Pin diameter required by shear [Eq. (6.2)]:

$$\text{shear load } V = \tfrac{400}{2} = 200 \text{ kips}$$

$$d = \sqrt{\frac{4V}{\pi F_v}} = \sqrt{\frac{4 \times 200}{\pi \times 14.5}} = 4.2 \text{ in.}$$

Pin diameter required by bearing:

$$d = \frac{P}{tF_p} = \frac{120}{0.78 \times 33} = 4.66 \text{ in.} \qquad \text{(flange of vertical strut)}$$

$$d = \frac{160}{1.00 \times 33} = 4.75 \text{ in.} \qquad \text{(on chord)}$$

$$d = \frac{200}{1.25 \times 33} = 4.85 \text{ in.} \qquad \text{(on eyebar)}$$

Use pin diameter $d = 5.0$ in.

Note: Attachment of the pin plate to the channel web is deferred to Problem 6.5.

6.4 WELDED CONNECTIONS

Structural welds are usually made either by the manual shielded-metal-arc process or by the submerged-arc process, the latter being especially suited to automatic shop welding of built-up members with controlled positioning. In either process the heat of an electric arc simultaneously melts the welding electrode and the adjacent steel in the parts being joined. The electrode is deposited in the weld as filler metal. The wide adoption of welding in recent years has required improved control of steel chemistry in order to provide steels that are "weldable," that is, steels that can be joined together with sound metal, of adequate strength and ductility, and with minimal metallurgical damage to adjacent parent metal.

In the shielded-metal-arc process, pictured in Fig. 6.4, the electrode coating creates a gaseous shield that protects the molten weld metal from the atmosphere. In the submerged-arc process, the arc occurs underneath a previously deposited fusible powdered flux that blankets the welding zone, and the bare electrode usually is fed automatically from a reel of wire.

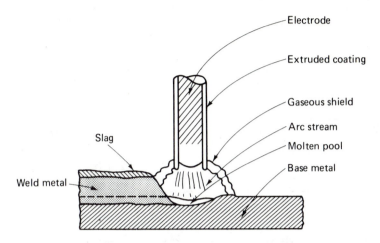

Electrode

Extruded coating

Gaseous shield

Arc stream

Molten pool

Base metal

Slag

Weld metal

Fig. 6.4 Shielded-metal-arc welding process.

The adoption of rules governing the qualification of welders and welding processes together with quality control of all materials have brought welding to the point where it is today permitted for practically all steel fabrication for both shop and field connections. Welding offers many advantages, which may be briefly outlined as follows:

1. Simplicity of design details, efficiency, and minimum weight are achieved because welding provides the most direct transfer of stress from one member to another.
2. Fabrication costs are reduced because fewer parts are handled and operations such as punching, reaming, and drilling are eliminated.
3. There is a saving in weight in main tension members since there is no reduction in area due to rivet and bolt holes. Additional saving is also achieved because of the fewer connecting parts required.
4. Welding provides the only plate-joining procedure that is inherently air- and watertight and hence is ideal for water and oil storage tanks, ships, and so forth.
5. Welding permits the use of fluidly changing lines that enhance the structural and architectural appearance, as well as reduce stress concentrations due to local discontinuities.
6. Simple fabrication becomes practicable for those joints in which a member is joined to a curved or sloping surface, such as structural pipe connections.
7. Welding simplifies the strengthening and repair of existing riveted or welded structures.

The two most common types of welds are fillet welds and groove welds. *Fillet welds* are used to attach a plate to another plate or member in either a

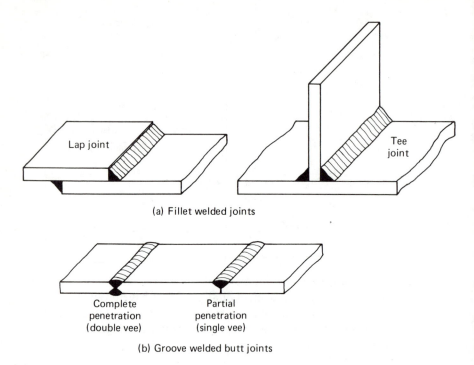

(a) Fillet welded joints

(b) Groove welded butt joints

Fig. 6.5 Types of welded joints.

parallel (lapped) or protruding (tee) position, as shown in Figs. 6.5(a). *Groove welds*, as shown in Fig. 6.5(b), retain the continuity of plate elements that are butt joined along their edges. Groove welds require special edge preparation and careful fit up, and when welded from both sides, or from one side with a backup strip on the far side, they may be said to achieve *complete penetration* and may be stressed as much as the weakest piece that has been joined. Incomplete-penetration groove welds are used only when the plates are not required to be fully stressed and full continuity is not required. Complete-penetration groove welds are also used for corner or tee joints when full plate development is required.

AISCS, Table 1.5.3, indicates that allowable stresses in compression, tension, and shear for full-penetration groove welds are the same as for the base metal. Partial-penetration groove welds are permitted full stress effectiveness only for compression normal to the effective throat. They may be stressed in tension on the effective throat at a reduced allowable stress equal to that allowed in shear.

Standard weld symbols to be used in steel detailing are shown in the AISCM with their correct use illustrated in many sketch details.

Fillet welds are more easily made than groove welds because of the greater fit-up tolerances allowed. As shown in Fig. 6.5, fillet welds are commonly used in connecting lapped plates or projecting plate elements to another plate or member. The allowable force transmitted by a unit length of fillet weld is equal to the product of the *effective throat dimension* multiplied by the allowable *shear* stress, as listed in Table 1.5.3, for the electrode and base metal specified therein. The effective throat dimension T_e is illustrated in Fig. 6.6, where w is the nominal weld size and t the throat dimension.

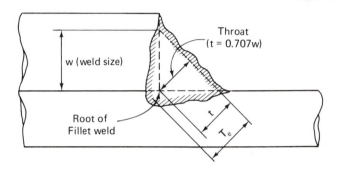

Fig. 6.6 Fillet-weld nomenclature.

When the welded faces of the joined parts are at 90°, as shown,

$$t = 0.707w$$

When the manual shielded-metal-arc process is used, the effective throat is equal to the dimension t. When the submerged-metal-arc process is used, greater heat input produces a deeper penetration and an effectively greater throat dimension is allowed. Thus for the submerged-arc process, AISCS, Sec. 1.14.6.2, permits the effective throat, T_e, to be taken as equal to the weld size, w, when w is $\frac{3}{8}$ in. or less, and equal to $t + 0.11$ when w is greater than $\frac{3}{8}$ in.

Table 1.5.3 lists allowable stresses for various electrode and base-metal combinations for fillet welds. The *resultant stress* on the effective throat is taken to be *equivalent* to shear stress in determining the allowable force on a unit length of weld. This concept is illustrated in Fig. 6.7, where the z axis is arbitrarily shown along the fillet weld throat and the x and y axes are in the surface planes of the tee joint, as shown in Fig. 6.7(b). The region above lines A–A–A has been removed in the pictorial representation of Fig. 6.7(a), in which the section through the fillet throat is exposed. The resultant stress f_r, having components f_x, f_y, and f_z, must be kept below the permissible stress values listed in AISCS, Table 1.5.3. It may be noted that f_z is a pure shear component, directed along the weld, and that f_x and f_y include both shear

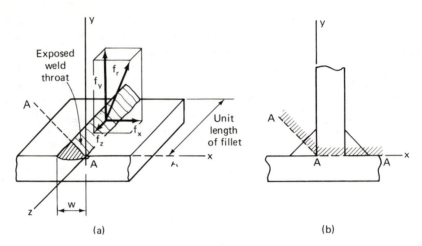

Fig. 6.7 Stress components and resultant stress on fillet-weld throat.

and normal components. The stress resultant, f_r, is

$$f_r = \sqrt{f_x^2 + f_y^2 + f_z^2} \tag{6.3}$$

The x, y, and z axes may be oriented in any arbitrary direction. The use of the stress resultant as a strength criterion has been shown to be adequate by many tests of welded connections, including those that are eccentrically loaded.

Fillet welds are specified on drawings and in design calculations by their size, w, which varies from $\frac{3}{16}$ to $\frac{1}{2}$ in. in $\frac{1}{16}$-in. increments, and which varies in $\frac{1}{8}$-in. increments for sizes greater than $\frac{1}{2}$ in. Weld sizes of $\frac{3}{16}$, $\frac{1}{4}$, and $\frac{5}{16}$ in. are favored because they can be made by a single pass of the electrode. The amount of filler metal increases as the square of the weld size. Thus time and cost of welding increase disproportionately as the weld size increases.

There are other limitations on fillet-weld size. A small weld at the edge of a thick plate is chilled rapidly, which has an embrittling effect, and it may crack as it tries to shrink when it cools while being restrained by the heavy plate. Minimum permitted sizes of fillet welds, in reference to the thicker plate joined, are listed in Table 1.17.2A of the AISCS. Fillet welds along the edges of plates thicker than $\frac{1}{4}$ in. are also limited as to *maximum* size, which can be no greater than the plate thickness minus $\frac{1}{16}$ in. (AISCS, Sec. 1.17.3). When a fillet weld is terminated, small sections near the ends are not fully effective. Rules of design covering this and other related topics are covered in AISCS, Secs. 1.17.4. to 1.17.8.

Allowable weld strength values per lineal inch are most conveniently quoted in terms of the weld size, w, even though determined by the resultant stress on the throat. Weld sizes are called for in multiple values of $\frac{1}{16}$ in. It is therefore convenient in design to use the strength value of a $\frac{1}{16}$-in. weld as a basic unit. Thus, if F_v is the allowable shear stress from AISCS, Table 1.5.3, the allowable shear strength (kips per inch of weld length and per $\frac{1}{16}$ in. of fillet weld size) will be termed $\bar{q}$, where

$$\bar{q} = \tfrac{1}{16}(0.707)F_v$$

Letting $N =$ the number of $\frac{1}{16}$ in. in a weld (for example, $N = 4$ for a $\frac{1}{4}$-in. weld), the allowable shear in kips per inch of fillet weld is termed q_a, where

$$q_a = N\bar{q} \qquad \text{where } N = 16w$$

$$w = \frac{t}{.707}$$

For submerged-arc welds, the greater effective throat was previously defined (AISCS, Sec. 1.14.6.2). Allowable shears per $\frac{1}{16}$ in. of weld size are equal to $F_v/16$ for welds of $\frac{3}{8}$ in. or less. For weld sizes over $\frac{3}{8}$ in. the allowable shear stresses are the same as for the metal-arc process, plus a fixed bonus of $0.11F_v$. These values are listed and explained in Table 6.1 for the six different allowable stress levels established in AISCS, Table 1.5.3.

Table 6.1

Allowable Shear (kips/in.) of Fillet Welds, $\bar{q}$,
per $\frac{1}{16}$ in. of Weld Size for Various Permissible
Stress Levels from AISCS, Table 1.5.3

	Metal-Arc Electrode					
	E60	*E70*	*E80*	*E90*	*E100*	*E110*
Allowable shear stress, F_v (ksi), on effective throat of fillet weld	18.0	21.0	24.0	27.0	30.0	33.0
Allowable weld shear ($\bar{q}$) (kips/in.) per $\frac{1}{16}$ in. of weld size						
Metal-arc process	0.80	0.93	1.06	1.19	1.33	1.46
Submerged-arc size $\frac{3}{8}$ in. or less	1.12	1.31	1.50	1.69	1.88	2.06
Submerged-arc bonus† for size over $\frac{3}{8}$ in., but use metal-arc's allowable shear	1.98	2.31	2.64	2.97	3.30	3.63

†For example, for a $\frac{1}{2}$-in. submerged-arc weld and for electrode E80, $q_a = 8 \times 1.06 + 2.64 = 11.12$ kips/in.

The designer has two choices in laying out a suitable concentrically loaded fillet weld to transmit a given connection load P:

1. Select a weld size, determine the weld value q_a in kips per lineal inch of weld, and then determine the total length l of weld required:

$$l = \frac{P}{q_a}$$

In using Table 6.1, $q_a = N\bar{q}$ for the metal-arc process welds and for submerged-arc welds of size $\frac{3}{8}$ in. or less. For submerged-arc weld sizes greater than $\frac{3}{8}$ in., use the footnote to Table 6.1.

2. Alternatively, if a particular weld length is suggested by the geometry of the connected parts, the required weld shear force is determined:

$$q = \frac{P}{l}$$

The required weld size (N) in sixteenths may then be determined by dividing the calculated shear force per inch (q) by $\bar{q}$ as listed in Table 6.1. This procedure must be modified for submerged arc welds of size greater than $\frac{3}{8}$ in. by first subtracting the tabulated bonus and then dividing the calculated remaining force by the *metal-arc electrode* listing for $\bar{q}$.

In designing a double fillet-welded tee joint [Fig. 6.5(a)] the allowable shear force per inch of two fillet welds may exceed the allowable shear force per inch of stem of tee, in which case the latter would control the design and impose an effective upper limit on the useful size of the fillet welds. Fillet-welded tee connections to a single plate should always be welded on both sides because of the weakness in bending and susceptibility to shipping damage of a single fillet weld.

For single or double angles under static tension load, the AISCS, Sec. 1.15.3, does not require that fillet welds in end connections be disposed so as to balance the forces about the neutral axis of the member. However, for members in compression, or subject to repeated stress variation, it is recommended that the fillet welds be so placed as to balance the forces about the neutral axis and eliminate eccentricity. Two common cases occur, as illustrated in Fig. 6.8: case 1, comprised of two longitudinal welds, and case 2, with a transverse weld added.

Case 1

Assume all fillet welds to be the same size. Then the equilibrium conditions of forces and moments about the center-of-gravity axis of the angle member are given, respectively, as follows:

$$l = l_1 + l_2 = \frac{P}{q_a} \tag{6.4}$$

$$c_1(l_1 q_a) = c_2(l_2 q_a) \tag{6.5}$$

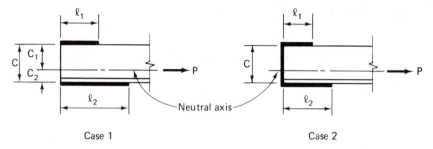

Case 1   Case 2

Fig. 6.8 Balancing fillet welds for angle connections.

from which is obtained

$$l_1 = \frac{c_2}{c} l \tag{6.6}$$

$$l_2 = \frac{c_1}{c} l \tag{6.7}$$

where

l_1, l_2 = required lengths of fillet welds

l = required total length of fillet welds

c_1, c_2 = distances from neutral axis to extreme fibers of the angle

P = axial load

q_a = permissible shear strength of fillet weld

Case 2

If a weld of length c is provided along the end of the angle, the equilibrium conditions of forces and moments again lead to the following expressions:

$$l = l_1 + l_2 + c = \frac{P}{q_a} \tag{6.8}$$

$$c_1 l_1 q_a + c\left(\frac{c}{2} - c_2\right) q_a = c_2 l_2 q_a \tag{6.9}$$

Then

$$l_1 = \frac{c_2}{c} l - \frac{c}{2} \tag{6.10}$$

$$l_2 = \frac{c_1}{c} l - \frac{c}{2} \tag{6.11}$$

Concentrically loaded welded connection design will now be illustrated by means of Examples 6.5, 6.6, and 6.7.

Example 6.5 *Lap Joint (Welded)*

Similar to Example 6.1, except that plates are PL $\frac{1}{2}$ × 8 loaded to 50-kip tension. Determine the transverse weld size required, using A36 steel and E70 electrodes.

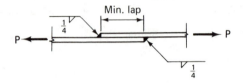

Solution

$$\text{minimum lap} = 5 \times \text{thickness of thinnest plate} \qquad \text{(AISCS, Sec. 1.17.6)}$$
$$= 5 \times \tfrac{1}{2} = 2.5 \text{ in.}$$

total length of weld = approx. 8 × 2 = 16 in.

Required weld unit shear force:

$$q = \tfrac{50}{16} = 3.13 \text{ kips/in.}$$

Required weld size:

$$N = \frac{q}{\bar{q}} = \frac{3.13}{0.93} = 3.4$$

$$w = \frac{N}{16} = \frac{3.4}{16}$$

Hence use $\frac{1}{4}$-in. fillet welds.

Check the minimum permitted weld size, AISCS, Table 1.17.2A:

$$\text{minimum weld size} = \tfrac{3}{16} \text{ in.} < \tfrac{1}{4} \qquad \text{OK}$$

Example 6.6 *Butt Joint with Groove Weld*

A PL $\frac{1}{2}$ × 12 carrying a tensile force of 125 kips is to be spliced. Use a groove-welded butt splice, A36 steel, and E70 electrodes.

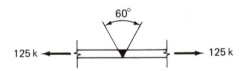

Solution

Actually, the only consideration here is to join the two pieces with a complete-penetration groove weld using electrodes that provide a weld with equal or greater tensile strength than that of the plate. A single vee complete-penetration groove weld requires that the weld root on the underside must be gouged and back welded. This is only one of several groove welds that could be used. For example, if the underside is inaccessible to the welder, a backup strip must be tack welded to one of the pieces prior to fit up.

Example 6.7

Same as Example 6.3, except that fasteners are replaced by welds. The bolted tee may now be replaced by a single $\frac{5}{8}$-in. plate† that is welded directly to the center of the column flange using a double-bevel full-penetration groove weld. Fillet welds attaching the angles to the plate should be arranged so as to balance the forces about the neutral axis of the connection of the double angles to eliminate any eccentricities.

Solution

A. *Connection A:* Assume that no weld is placed along the end of angles. Use maximum allowable weld size to connect angles to bracket. Then $w = \frac{1}{2} - \frac{1}{16} = \frac{7}{16}$-in. weld (AISCS, Sec. 1.17.3). The permissible strength of $\frac{7}{16}$-in. fillet weld (use E70 electrodes) is

$$q_a = 0.707wF_v \qquad 70(.3)$$
$$= 0.707 \times \tfrac{7}{16} \times 21 = 6.5 \text{ kips/in.}$$

(or use Table 6.1: $q_a = 7 \times 0.93 = 6.5$ kips/in.). Here F_v is the permissible shear stress of weld (AISCS, Table 1.5.3).

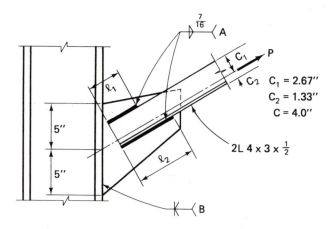

†*Note:* As an alternative to the use of the $\frac{5}{8}$-in. plate, the WT 15 × 58 could again be used, as in Example 6.3, but the simplicity inherent to welded design would then be lost. It may be noted, however, that the protruding plate attached permanently in the shop to the column is a nuisance and subject to damage in shipment.

The required total weld length is

$$l = \frac{P}{q_a} = \frac{120/2}{6.5} = 9.23 \text{ in.}$$

$$l_1 = \frac{c_2}{c}l = \frac{1.33}{4}9.23 = 3.07 \text{ in.} \qquad [\text{Eq. (6.6)}]$$

$$l_2 = \frac{c_1}{c}l = \frac{2.67}{4}9.23 = 6.17 \text{ in.} \qquad [\text{Eq. (6.7)}]$$

Use $l_1 = 3\frac{1}{4}$ in.; $l_2 = 6\frac{1}{2}$ in.

B. *Connection B:* According to AISCS, Table 1.5.3, the allowable stresses in tension and shear in a full penetration groove weld are the same as for the base metal. The AISCS does not deal explicitly with the problem of combined shear and tension in groove welds, but an adequate design may be obtained by limiting the maximum principal tensile stress to the allowable stress in tension. As discussed later in Chapter 10, this may be accomplished by a closely approximate formula [Eq. (10.13)], which simply reduces the allowable tension stress in the direction of applied tensile force to a value F_{rt}:

$$F_{rt} = \left[1 - \left(\frac{f_v}{F_t}\right)^2\right]F_t$$

The tension and shear components of the applied force in the groove weld, designated respectively as T and V, are (Example 6.3)

$$T = 104 \text{ kips}$$

$$V = 60 \text{ kips}$$

Try $l = 10$ in. (double-bevel full penetration groove weld):

$$f_v = \frac{V}{tl} = \frac{60}{\frac{5}{8} \times 10} = 9.6 \text{ ksi}$$

$$f_t = \frac{T}{tl} = \frac{104}{\frac{5}{8} \times 10} = 16.6 \text{ ksi}$$

By Eq. (10.13),

$$F_{rt} = \left[1 - \left(\frac{9.6}{22}\right)^2\right]22 = 17.8 \text{ ksi} > f_t \qquad \text{OK}$$

6.5 ECCENTRICALLY LOADED CONNECTIONS

Previous design examples in this chapter have been limited to concentrically loaded connections. Concentricity is always desirable, but eccentrically loaded connections are sometimes required. Riveted, bolted, and welded connections will be considered.

When eccentricity of load imposes only shear force on riveted and bolted fasteners, with no change of initial tension, the assumption may be made that the eccentric load may be replaced by an equivalent force and couple acting at the centroid of the fastener group. If, on the other hand, both shear and tension are induced by load eccentricity, the design will require consideration of the following:

1. Reduced allowable tension stress in fasteners for bearing-type riveted or bolted connections, or
2. Reduced allowable shear stress in friction-type fasteners using high-strength bolts.

The foregoing effects were considered in Section 6.2.

Compact fillet-welded connections between heavy parts may be designed simply by providing a weld size adequate to resist the maximum resultant force per lineal inch of weld due to the combined effect of applied force and the moment induced by eccentricity.

Initial attention is now given to riveted or bolted shear connections that produce no tensile force change in the fasteners, as shown by the two-plate bracket bolted to the flanges of a W column in Fig. 6.9(a). The load P may be vertical, or inclined as shown, and there is an eccentric moment of magnitude Pe acting at the centroid of the fastener group, as shown in Fig.6.9(b). In this illustration, because of the double symmetry of the fastener pattern, the centroid is readily located by inspection. It is convenient to replace P by its components P_x and P_y, as shown in Fig. 6.9(b). The problem is to determine the maximum resultant shear force on the particular most-stressed fastener, which usually can be located by elementary considerations.

In Fig. 6.9(c), R_{AP} represents the resultant force due to applied load P acting on any *particular* bolt A. As assumed for concentric connections, $R_{AP} = P/n$, where n is the total number of fasteners. R_{Am} is the resultant force due to moment, assumed to be proportional to the radial distance r_A from centroid O, and to act normal to r_A. The total shear force on the fastener is shown as R_A, the resultant of R_{AP} and R_{Am}, and is shown graphically as the diagonal of the force parallelogram. To avoid the determination of angles and the use of unwieldy arithmetic, it is convenient to break down each of these resultant forces into its x and y components.

We let R_{xp} and R_{yp} represent the shear force components in a single fastener due to P, respectively, in the x and y directions. Let R_{xm} and R_{ym} be the shear force components in the same fastener due to M, respectively, in the x and y directions.

Consider P to be replaced by its components, P_x and P_y:

$$R_{xp} = \frac{P_x}{n} \quad \text{and} \quad R_{yp} = \frac{P_y}{n} \tag{6.12}$$

In Fig. 6.9(c) let I be *any* fastener at radial distance r_i from the centroid of the fastener group (O), and R_{im} the force due to the eccentric moment.

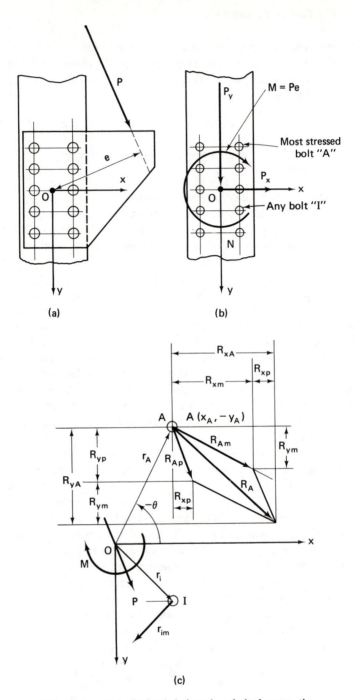

Fig. 6.9 Eccentrically loaded riveted or bolted connection.

The contribution of fastener I to the moment resistance is equal to $R_{im}r_i$, and the total moment is equal to the sum of these individual contributions for all n fasteners:

$$M = \sum R_{im}r_i \tag{6.13}$$

As assumed,

$$\frac{R_{Am}}{R_{im}} = \frac{r_A}{r_i} \quad \text{or} \quad R_{im} = \frac{R_{Am}r_i}{r_A}$$

Substituting this value of R_{im} into Eq (6.13),

$$M = \frac{R_{Am}}{r_A} \sum r_i^2 \tag{6.14}$$

or the force R_{Am}, due to moment, on any particular fastener is equal to

$$R_{Am} = \frac{Mr_A}{\sum r_i^2} \tag{6.15}$$

Breaking R_{Am} into its x and y components, and noting, by similar triangles, that

$$\frac{R_{xm}}{R_{Am}} = -\frac{y_A}{r_A} \qquad \frac{R_{ym}}{R_{Am}} = \frac{x_A}{r_A}$$

then

$$R_{xm} = -\frac{y_A R_{Am}}{r_A} = -\frac{My_A}{\sum r_i^2} \tag{6.16a}$$

$$R_{ym} = \frac{x_A R_{Am}}{r_A} = \frac{Mx_A}{\sum r_i^2} \tag{6.16b}$$

In calculating $\sum r_i^2$, it is convenient to use x and y component distances:

$$\sum r_i^2 = \sum x_i^2 + \sum y_i^2 \tag{6.17}$$

The x and y components of total shear force on fastener A due to both applied load and moment are

$$R_{xA} = R_{xp} + R_{xm} = \frac{P_x}{n} - \frac{My_A}{\sum r_i^2} \tag{6.18a}$$

$$R_{yA} = R_{yp} + R_{ym} = \frac{P_y}{n} + \frac{Mx_A}{\sum r_i^2} \tag{6.18b}$$

Finally, the resultant force on the most-stressed fastener is determined:

$$R_A = \sqrt{R_{xA}^2 + R_{yA}^2} \tag{6.19}$$

and the connection design is adequate if

$$\text{max. } R_A \leq \text{allowable shear force}$$

In Fig. 6.9 it is obvious by means of the following reasoning that the most-stressed bolt is at A:

1. R_{xp} and R_{yp} act in the positive x and y directions.
2. A is one of four locations where R_{lm} is maximum.
3. A is the only bolt for which both R_{xm} and R_{ym} act in the same direction and thereby increase the magnitude of R_{xp} and R_{yp}.

Turning now from the riveted or bolted connection to the eccentrically loaded fillet-welded connection, the analysis is essentially similar and will not be developed in as much detail. In place of any single fastener, one now considers a differential length of fillet weld dl, and in place of $\sum r_i^2$ (which is the polar moment of inertia of the pattern of unit bolt areas), we substitute for a unit width of fillet-weld throat,

$$\int r_i^2 \, dl = I_z$$

The calculation of I_z is illustrated in Example 6.9 which makes use of the relationship $I_z = I_x + I_y$.

Figure 6.10 shows an eccentrically loaded plate bracket welded to a column flange to provide a connection similar to the bolted bracket in Fig. 6.9. The same-size fillet weld is assumed along the three edges of the bracket

Fig. 6.10 Eccentrically loaded welded connection.

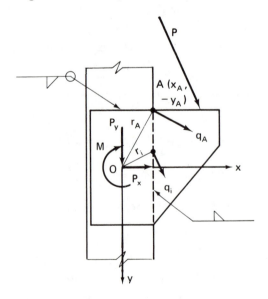

plate and one hidden edge of the column flange. Let the component forces in the x and y directions per unit length of weld at location A and due to applied force P be

$$q_{xp} = \frac{P_x}{l} \tag{6.20a}$$

$$q_{yp} = \frac{P_y}{l} \tag{6.20b}$$

l represents the total length of fillet weld measured at the root location.

The component shear forces per unit length at the same location due to moment can be shown to be

$$q_{xm} = -\frac{My_A}{I_z} \tag{6.21a}$$

$$q_{ym} = \frac{Mx_A}{I_z} \tag{6.21b}$$

I_z is the moment of inertia of the unit width weld about the z axis (through O in Fig. 6.10) and normal to the xy plane. As in the case of the bolted connection,

$$q_x = q_{xp} + q_{xm} = \frac{P_x}{l} - \frac{My_A}{I_z} \tag{6.22a}$$

$$q_y = q_{yp} + q_{ym} = \frac{P_y}{l} + \frac{Mx_A}{I_z} \tag{6.22b}$$

Thus the resultant force per unit length of weld at point A—the determining factor in selecting the weld size—is

$$q_A = \sqrt{q_x^2 + q_y^2} \tag{6.23}$$

The foregoing procedure may be readily extended to a general three-dimensional fillet-welded joint with the addition of a q_z force component.

Riveted and bolted connections in which eccentricity induces change in fastener tension will be considered in Section 6.7. Examples 6.8 and 6.9 illustrate the design of shear-type connections using high-strength bolts and welds, respectively.

Example 6.8 *Eccentric Loads on High-Strength Bolted Bracket*

Design a bracket connected to the faces of a W 14 × 193 column to support a girder reaction of 130 kips applied eccentrically, as shown. Using four vertical lines of bolts in each face of the column, determine the number and size of A490 high-strength bolts in friction-type connection in accordance with the AISCS.

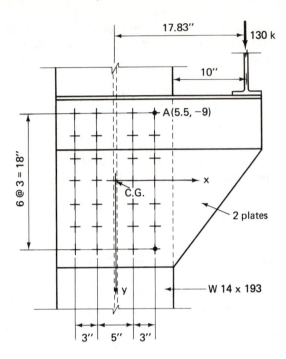

Solution

Assume seven bolts with 3-in. spacing in each vertical line. The centroid of the bolt group is at point O from symmetry. Shear stresses in bolts shall be in accordance with AISCS, Table 1.5.2.1.

$$P_x = 0$$

$$P_y = \frac{130}{2} = 65 \text{ kips}$$

$$M_z = 65 \times 17.83 = 1159 \text{ kip-in.} \quad \text{(in each plate)}$$

$$n = 7 \times 4 = 28 \text{ bolts}$$

$$\sum r_i^2 = \sum (x_i^2 + y_i^2) = 14(5.5^2 + 2.5^2) + 8(9^2 + 6^2 + 3^2) = 1519 \text{ in.}^2$$

By inspection, the bolt A in the right-hand upper corner (or right-hand lower corner) is the most-stressed bolt. Compute the resultant shear force R on the upper-right-corner bolt where $x = 5.5$ in. and $y = -9$ in.; then according to Eqs. (6.12), (6.16), (6.17), and (6.19),

$$R_x = \frac{P_x}{n} - \frac{M_z y}{\sum r_i^2} = 0 - \frac{1159 \times (-9)}{1519} = 6.87 \text{ kips/bolt} \longrightarrow \quad (+ \text{ direction})$$

$$R_y = \frac{P_y}{n} + \frac{M_z x}{\sum r_i^2} = \frac{65}{28} + \frac{1159 \times 5.5}{1519} = 6.52 \text{ kips/bolt} \downarrow \quad (+ \text{ direction})$$

$$R_A = \sqrt{R_x^2 + R_y^2} = \sqrt{6.87^2 + 6.52^2} = 9.47 \text{ kips/bolt}$$

Try $\frac{3}{4}$-in. bolts; the allowable shear force in single shear is

$$A_b F_v = 0.44 \times 22 = 9.68 \text{ kips/bolt} > R_{\max} \qquad \text{OK}$$

where

A_b = nominal body area of a bolt
F_v = allowable shear stress on bolts (see AISCS, Table 1.5.2.1)

Use 28 A490 high-strength bolts of $\frac{3}{4}$-in. diameter.

Example 6.9

For a single plate bracket-welded to the face of a column, as shown in the figure, determine the size of the fillet welds required. Use E70 electrodes and A36 steel.

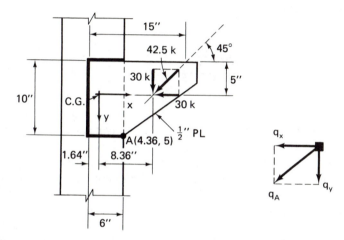

Solution

Find the center of gravity of the weld root line. Vertically, by inspection, it is 5 in. from the top or bottom of the plate, because of symmetry. Horizontally, $(2 \times 6 \times 3)/22 = 1.64$ in.

Design loads acting at C. G. of weld group:

$$P_x = -42.5 \cos 45° = -30 \text{ kips}$$
$$P_y = 42.5 \sin 45° = 30 \text{ kips}$$
$$P_z = 0$$
$$M_x = M_y = 0$$
$$M_z = 30 \times 8.36 = 251 \text{ kip-in.}$$

Moment of inertia:

$$I_x = 2 \times 6 \times 5^2 + \tfrac{1}{12} \times 10^3 = 384 \text{ in.}^4/\text{in.}$$
$$I_y = 10 \times 1.64^2 + 2[\tfrac{1}{3} \times (1.64^3 + 4.36^3)] = 85 \text{ in.}^4/\text{in.}$$
$$I_z = I_x + I_y = 384 + 85 = 469 \text{ in.}^4/\text{in.}$$

Note the shearing stresses on the sketch, which indicate that the lower right edge (where $x = 4.36$, $y = -5$ and all stress components are additive) is the most highly stressed:

$$q_x = \frac{P_x}{l} - \frac{M_z y}{I_z} = -\frac{30}{22} - \frac{251 \times 5}{469} = -1.36 - 2.68 = -4.04 \text{ kips/in.} \longleftarrow$$

$$q_y = \frac{P_y}{l} + \frac{M_z x}{I_z} = \frac{30}{22} + \frac{251 \times 4.36}{469} = 1.36 + 2.33 = 3.69 \text{ kips /in.} \downarrow$$

$$q_A = \sqrt{q_x^2 + q_y^2}$$
$$= \sqrt{4.04^2 + 3.69^2} = 5.47 \text{ kips/in.}$$

From Table 6.1, E70 weld value $\bar{q} = 0.93$ kip/in. Weld size required in $\frac{1}{16}s = N = 5.47/0.93 = 5.9$. Use $\frac{3}{8}$ in.

6.6 SHEAR CONNECTIONS FOR BUILDING FRAMES

A variety of beam-to-column or beam-to-girder connections is available to support simple beam reactions. They are purposely made flexible with regard to rotation between the ends of the beam and the column or girder. These are designated by AISCS as Type 2 connections and are used in structures for which lateral forces do not need to be considered, or where other bents in the building resist wind and seismic forces by frame action, truss framing, or shear walls. Flexible connections for reactions may involve attachment only to the beam web, as shown in Fig. 6.11, or may consist of top and bottom angles, respectively designated by AISCS as "framed beam connections" and "seated beam connections." They may include rivets, bolts, or welds, alone or in combination. The AISCM provides descriptive and tabular design information covering the most commonly used types. These connections do develop a certain amount of moment, which may amount to 10% of the full fixed end moment or even more. However, these moments are dis-

Fig. 6.11 Flexible framing angle connection.

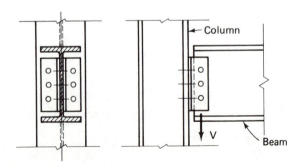

regarded in design. A nominal end clearance or "setback" between beam end and column of $\frac{1}{2}$-in. is assumed, but connections are designed for a $\frac{3}{4}$-in. setback to allow for possible underrun in beam length.

In the case of the riveted or bolted stiffened seat angle, the fasteners that attach the angles to the column or girder are in combined shear and tension due to the eccentricity of the applied load. The bending moment is transmitted through the connection by tension in the upper fasteners and by bearing pressure in the lower part between the angles and the column, as shown by the shaded area in Fig. 6.12. The equivalent effective area in transmitting the bending moment is shown in Fig. 6.12(b) in which area ac_1 equals the area of the eight fasteners and whose stress distribution, due to moment, is shown in Fig. 6.12(c). Then

$$a = \frac{mA}{p} \tag{6.24}$$

$$c_1 = \frac{\sqrt{b}}{\sqrt{a} + \sqrt{b}} h \tag{6.25}$$

$$R_{\max} = \frac{M}{I}(c_1 - e)A \tag{6.26}$$

where

$a =$ width of the equivalent fastener area
$p =$ pitch of the fasteners
$m =$ number of the fasteners per horizontal row
$A =$ cross-sectional area of a fastener
$c_1, c_2 =$ distance between neutral axis and extreme fibers
$b =$ width of the framing angles or tee
$h =$ height of the framing angles or tee
$R_{\max} =$ maximum tensile load in a fastener
$I =$ moment of inertia $= (ac_1^3 + bc_2^3)/3$
$e =$ edge distance between the maximum tensile load of the fastener and the edge of the framing angles

The procedure illustrated in Fig. 6.12 applies only to the specialized case in which the vertical spacing of the fasteners is uniform. However, this happens very frequently in practice, and the equations that were developed are very simple to use.

If the vertical spacing of the fasteners varies, we would have a varying thickness (a) that could not be used in these equations. Therefore, we must use another method, in which we consider the moments of the individual fasteners about the neutral axis. The neutral axis usually lies between one-sixth to one-seventh of the length of the connection (h) from the bottom of the connection. This assumption is only used to estimate the number of fasteners in tension (above the neutral axis). An equation is then written in which the moment of the compression area about the neutral axis is equated

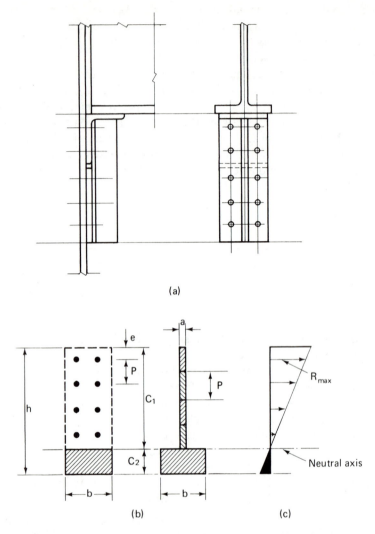

Fig. 6.12 Seat angle connection with fasteners subjected to tension: (a) actual stress portions; (b) equivalent cross section; (c) stress distribution.

to the moment of the areas of the tension fasteners about the neutral axis, since these must be equal. The resulting answer will indicate whether we have made a correct assumption.

Once the neutral axis is located, we can calculate the moment of inertia. Next, using the flexure formula, we can determine the tensile stress in the critical fasteners and the compression stress on the extreme fiber of the connection.

$$f_t = \frac{My}{I}$$

An application of the preceding procedure is illustrated in Example 6.11. Other connection design principles required in the examples of beam shear connections that follow are based on procedures that have been covered previously in this chapter.

Example 6.10 *Seated Beam Connection (Bolted or Riveted)*

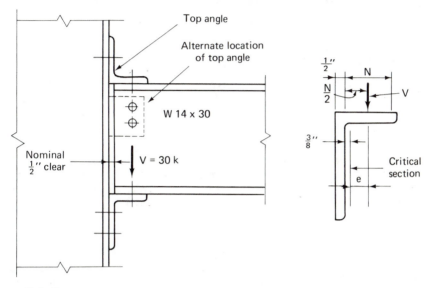

Solution

For W 14 × 30, $b_f = 6.73$ in., $t_w = 0.27$ in., and $k = 0.875$. The length of bearing required is

$$N + k = \frac{R}{0.75 F_y t_w} \qquad \text{[AISCS Formula (1.10-9)]}$$

Then

$$N = \frac{30}{0.75 \times 36 \times 0.27} - 0.875 = 3.24 \text{ in.}$$

Use a 4-in. leg, 8 in. long. Assume $\frac{5}{8}$-in. angle thickness (governed by bending); the distance from the back of the angle to the critical section in bending is $\frac{5}{8} + \frac{3}{8} = 1$ in. It is usually considered that the load is concentrated at the center of the required bearing (3.24 in. in this example). Allow 0.25 in. underrun.

$$\text{eccentricity} = 0.5 + 0.25 + \frac{3.24}{2} - 1.0 = 1.37 \text{ in.}$$

$$M = 30 \times 1.37 = 41.1 \text{ kip-in.}$$

$$F_b = \frac{6M}{bt^2} = \frac{6 \times 41.1}{8t^2}$$

Then

$$t^2 = 1.14 \quad (F_b = 27 \text{ ksi}) \qquad \text{and} \qquad t = 1.07 > \frac{5}{8} \quad \text{NG}$$

Try 10-in. length and $\frac{7}{8}$-in. thickness (eccentricity $= 1.12$ in. and $M = 33.6$ kip-in.), and by calculations paralleling the preceding, the required

$$t = 0.86 \text{ in.} < 0.875 \quad \text{OK}$$

Use L $6 \times 4 \times \frac{7}{8} \times 10$ in. long.

Alternative procedures for attachment to the column are:

A. *Riveted:* $\frac{3}{4}$-in. A502 grade 1 rivets good for $17.5 \times 0.4418 = 7.73$ kips (AISCS, Table 1.5.2.1):

$$n = \frac{30}{7.73} = 3.88$$

Use four rivets.

B. *Bolted:* $\frac{3}{4}$-in. A325 bolts in friction type connection with threads not excluded from shear plane, good for 7.73 kips (AISCS, Table 1.5.2.1):

$$n = \frac{30}{7.73} = 3.88$$

Use four bolts.

Connections of the beam to the seat angle, the beam to the top angle, and the top angle to the column can be made using two $\frac{3}{4}$-in.-diameter unfinished bolts. The top angle simply holds the beam in a vertical position and must be flexible to permit simple beam end rotation. Use L $4 \times 4 \times \frac{1}{4} \times 8$ in. long.

Example 6.11

Design a riveted stiffened seat connection for a W 18×46 and an end reaction of 40 kips. All material is A36 steel. Refer to the figure. Sketches (b) and (c) give additional details of the seat, and sketch (d) is for use in calculating rivet stress.

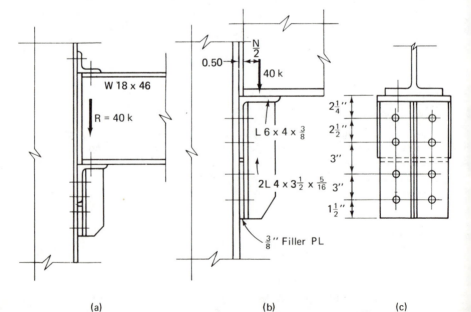

(a)   (b)   (c)

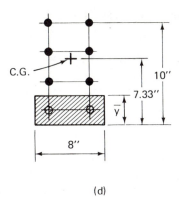

(d)

Solution

For W 18 × 46, $b_f = 6.06$ in., $t_w = 0.360$ in., and $k = 1.13$.

$$\text{req'd } N = \frac{40}{27 \times 0.36} - 1.13 = 2.99 < 3.25 \qquad \text{OK}$$

Assume a nominal setback of 0.5 in. plus 0.25 in. allowance for a beam length underrun. Eccentricities of load for the stiffened seats will be assumed as 0.75 in. plus half the remaining seat length.

Assumed eccentricity of end reaction:

$$e = 0.75 + \frac{3.25}{2} = 2.38 \text{ in.}$$

$$\text{moment} = 2.38 \times 40 = 95.2 \text{ kip-in.}$$

Try $\frac{3}{4}$-in. A502 grade 1 rivets.

Assume the neutral axis to be above the two lowest rivets and locate the center of gravity (c.g.) of the upper six rivets. From the bottom of the vertical angles to the c.g. the distance is

$$\frac{2 \times 4.5 + 2 \times 7.5 + 2 \times 10}{6} = 7.33 \text{ in.}$$

Locate the neutral axis [see sketch (d) and the discussion of Fig. 6.12].

$$8\bar{y}\frac{\bar{y}}{2} = 6 \times 0.442(7.33 - \bar{y})$$

$$4\bar{y}^2 + 2.65\bar{y} - 19.4 = 0$$

Solving,

$$\bar{y} = 1.90 \text{ in.}$$

$$I = 8 \times \frac{1.90^3}{3} + 2 \times 0.442(2.6^2 + 5.6^2 + 8.1^2) = 110 \text{ in.}^4$$

The tensile stress in the top rivets is $95.2 \times 8.1/110 = 7.01$ ksi. The shear stress

$$f_v = \frac{40}{8 \times 0.442} = 11.3 \text{ ksi}$$

The allowable tensile stress (AISCS, Sec. 1.6.3) is

$$F_t = 30.0 - 1.3 f_v = 30.0 - 1.3 \times 11.3 = 15.3 \text{ ksi} > 7.01 \qquad \text{OK}$$

The required bearing area of stiffeners is $40/32.4 = 1.23$ in.^{2}, and the area furnished (using L $4 \times 3\frac{1}{2} \times \frac{5}{16}$ stiffeners) is

$$(3.5 - 0.5)\tfrac{5}{16} \times 2 = 1.87 \text{ in.}^2. \qquad \text{OK}$$

(See AISCS, Appendix A, Table 1, for allowable bearing stress.)

Example 6.12

Redesign the stiffened seat of Example 6.11 as all welded, using A36 steel and E70 electrodes. Use a structural tee section, 4 in. in length, as shown.

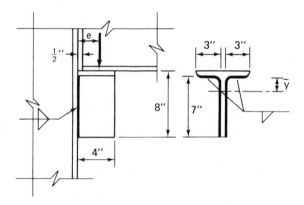

Solution

End eccentricity and moment are the same as for Example 6.11. Try WT 8 × 28.5 (which has a wider flange than the beam, to permit downhand fillet welds between beam and seat). Assume effective weld lengths as shown in the sketch, and locate the c.g. of the weld pattern (assume unit width of weld):

$$\bar{y} = \frac{3.5 \times 14.0}{20} = 2.45 \text{ in.}$$

$$I_x = 6 \times 2.45^2 + \tfrac{2}{3}(2.45^3 + 4.55^3) = 108.6 \text{ in.}^4/\text{in.}$$

Assume bearing of stem on column at bottom, making the resultant weld stress at the top the critical value. As a result of the eccentric moment,

$$q_H = \frac{95.2 \times 2.45}{108.6} = 2.15 \text{ kips/in.}$$

$$q_V = \tfrac{40}{20} = 2.0 \text{ kips/in.}$$

$$\text{resultant } q_R = \sqrt{2.15^2 + 2.0^2} = 2.94 \text{ kips/in.}$$

$$N = \frac{q_R}{q} = \text{weld size in } \tfrac{1}{16}s = \frac{2.94}{0.93} = 3.16 \qquad \text{(Table 6.1)}$$

Use $\frac{1}{4}$-in. fillet welds, which is the smallest size that meets the requirements of AISCS, Table 1.17.2A. Still greater size would be required if the column flange thickness exceeds $\frac{3}{4}$ in.

6.7 MOMENT-RESISTING CONNECTIONS

The two types of moment-resisting connections classified in AISCS, Sec. 1.2, as "rigid" (Type 1) and "semirigid" (Type 3) were discussed in Section 6.1.

Rigid connections are used in continuous-frame construction that resists lateral forces caused or induced by wind or earthquake. Rigid connections are also a requirement for frames that are proportioned according to plastic design, in which case the connections must be strong enough to develop complete yield moment at adjacent *plastic hinges*. Rigid connections are always advantageous if a building is loaded accidentally by explosive blast, earthquake, or high wind beyond its intended normal-use load. In such cases continuous-frame behavior, whether or not proportioned by plastic design, will provide a residual strength against ultimate collapse that may be a life-saving feature.

Semirigid connections are used in semicontinuous-frame construction primarily in office or apartment buildings of moderate height. The concept is intended to provide an economical balance between simple beam design, for which the maximum bending moment of $0.125wL^2$ is at the center, and fully continuous construction, for which the maximum moment of $0.083wL^2$ is at the ends in the fully fixed end condition. In a beam with semirigid connections, the end and center moments for uniform gravity load could, ideally, be balanced at $0.0625wL^2$, thus achieving savings in the weight of the beam. However, the careful balance that is required between strength and flexibility in semirigid connections is difficult to standardize and control; furthermore, the advent of plastic design and the development of fully continuous welded construction have lessened the economic advantage of semicontinuous construction. In current practice it is customary to label a connection as "semirigid" if it is designed for some specific moment capacity at any level less than that required to develop full continuity.

Figure 6.13 shows a riveted or bolted moment and shear connection that may be designed as either semirigid or rigid. It is a type that was much used prior to the advent of welding and it introduces the problem of prying action that increases bolt or rivet force. Another topic that will be covered briefly concerns the required strengthening of the column web against local deformation adjacent to the beam flanges.

In the connection shown in Fig. 6.13(a) the bending moment (M) is transmitted to the column flange by two tees (a) on the top and bottom flanges

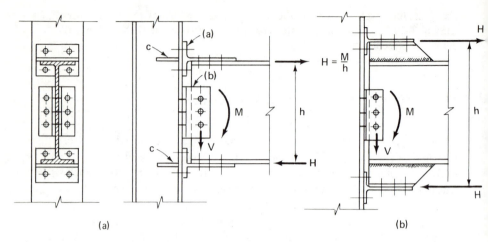

(a)   (b)

Fig. 6.13 Moment and shear connection.

of the beam by tension and compression loads (H), respectively. The shear load (V) is transmitted to the columns by the two angles (b) connected to the beam web. Then

$$H = \frac{M}{h}$$

where h is the depth of the beam. It should be noted that the moment capacity of this connection can be increased by increasing the distance between the tees, as shown in Fig 6.13(b). This greater distance will decrease the magnitude of force (H).

The top-flange moment connection shown in Fig. 6.13(a) transmits the applied moment to the column by means of fasteners acting in tension. Consequently, the connecting element (the tee section) is subjected to a bending stress and deforms as shown in Fig 6.14, thereby creating a prying action.

Fig. 6.14 Prying action in connection.

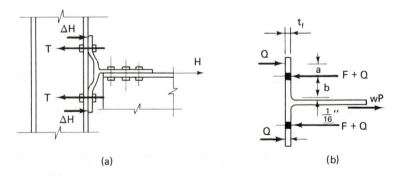

(a)   (b)

This prying action, in turn, creates a pressure (prying force ΔH) at the outer edge of the tee and thus adds directly to the total tension in the fasteners.

Research on the behavior of bolts in tee connections has led to the development of empirical formulas that approximate the prying force on the basis of simplifying and conservative assumptions. The prying force will be minimized and the desirable stiffness of the tee flange will be enhanced if the dimension b in Fig. 6.14(b) is kept as small as erection clearance requirements will permit.

The AISCM empirical formulas† for prying force Q are:

For connections using A325 bolts:

$$Q = \left[\frac{100bd_b^2 - 18wt_f^2}{70ad_b^2 + 21wt_f^2}\right]F \tag{6.27}$$

For connections using A490 bolts:

$$Q = \left[\frac{100bd_b^2 - 14wt_f^2}{62ad_b^2 + 21wt_f^2}\right]F \tag{6.28}$$

where

$P =$ allowable load on two angles or a structural tee, kips per lineal inch, as determined by stress due to bending, for allowable $F_b = 0.75F_y$
$Q =$ prying force per fastener, kips
$F =$ externally applied load per fastener ($wP/2$), kips
$w =$ length of flange tributary to each bolt, inches
$d_b =$ nominal bolt diameter, inches
$a =$ distance from fastener line to edge of flange, but not to exceed $2t_f$, inches
$t_f =$ flange thickness, inches
$b =$ distance from face of web to bolt line minus $\frac{1}{16}$ in., inches

By referring to the formulas it can be seen that when the flanges are thick and the gage lines are closely spaced, the prying action will be very small. In the preliminary selection of flange thickness it may be assumed that a point of inflection exists at the midlength of dimension b, in which case the flange moment per lineal inch is

$$M = \frac{Pb}{4} \tag{6.29}$$

In the final check of flange bending adequacy, the maximum moment may be assumed to be the greater of two values: at the fastener line, the total moment per fastener,

$$M_2 = Qa \tag{6.30}$$

†Check the latest AISC information for possible revision of these formulas.

and $\frac{1}{16}$ in. from the face of the tee stem,

$$M_1 = (F + Q)b - Q(a + b) = Fb - Qa \qquad (6.31)$$

The foregoing nomenclature corresponds to that used in the AISCM.

The design of column web stiffeners, [plates (c) in Fig. 6.13] is covered by AISCS, Sec. 1.15.5. Such stiffeners may be needed to prevent excessive local deformation in the column web and flange at locations near the top and bottom of the beam flanges where the moment resisting plates or tees are attached to the column flange.

Example 6.13 will illustrate the design of a semirigid beam-to-column connection with a moment resistance less than that required for full continuity. The procedure and formulas in AISCS, Sec. 1.15.5, will be applied and the effect of prying action evaluated by the conservative AISCM procedure and entered herein as Eq. (6.27).

Example 6.13 *Semirigid High-Strength Bolted Beam-to-Column Connection*

The connection is to be designed for a shear of 80 kips and a bending moment of 140 kip-feet using a design similar to that shown in Fig. 6.13(a) (see also the accompanying sketch). Forces are assumed to be due to maximum combined dead, live, and wind loads. Use A325 high-strength bolts and A36 steel.

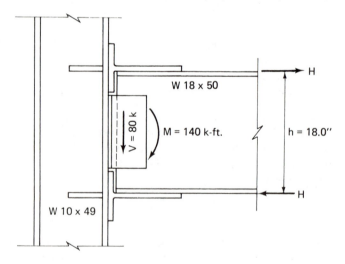

Since wind is included, allowable working stress levels may be increased by $\frac{1}{3}$ (AISCS, Sec. 1.5.6). A convenient way to accomplish the stress increase, in effect, is to multiply the design loads by $\frac{3}{4}$. Hence, for calculations at working stress levels, the connection will be designed for

$$M = \tfrac{3}{4} \times 140 = 105 \text{ kip-ft}$$

$$V = \tfrac{3}{4} \times 80 = 60 \text{ kips}$$

Solution

A. *Moment connectors:* The tension and compression forces transmitted by the tees, due to moment, at the top and bottom of the beam flanges are

$$H = \frac{M}{h} = \frac{105 \times 12}{18} = 70 \text{ kips}$$

1. *Required thickness of tee flanges:* Flange width of the W 18 × 50 beam is 7.5 in. Use the same length of tee section for the top and bottom connectors, selected initially on the basis of required flange thickness.

    Prying action will be neglected in the initial selection of the tees. Assuming a bolt gage of $g = 4$ in., we determine the tentative flange thickness on the assumption of full bending fixity at the bolt line. Bending moments in the flange are approximated as shown in the accompanying sketch.

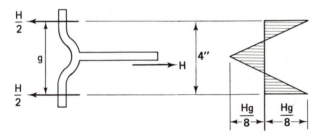

Moment in the tee flange:

$$m = \frac{Hg}{8} = \frac{70 \times 4}{8} = 35 \text{ kip-in.}$$

Required thickness:

$$t = \sqrt{\frac{6m}{bF_b}} = \sqrt{\frac{6 \times 35}{7.5 \times 27}} = 1.02 \text{ in.}$$

where

$$b = \text{length of tee}$$

$$F_b = \text{allowable stress in bending (AISCS, Sec. 1.5.1.4.3)}$$

Try a WT 9 × 59.5:

$$t_f = 1.06 \text{ in.} > 1.02$$

$$t_w = 0.655 \text{ in.}, \quad b_f = 11.265 \text{ in.}$$

2. *Tee web to beam flange*† (use $\frac{7}{8}$-in. A325 bolts in a bearing-type connection, threads excluded from shear plane):

$$F_v = 30 \text{ ksi} \qquad \text{(AISCS, Table 1.5.2.1)}$$

$$\text{number of bolts req'd} = \frac{70}{30 \times 0.601} = 3.88$$

Use four bolts, in two rows.

3. *Check the tensile capacity of the tee web:*
   a. *Gross section* ($F_t = 0.6F_y = 22 \text{ ksi}$):

$$H = 7.5 \times 0.655 \times 22 = 108 \text{ kips} > 70 \qquad \text{OK}$$

   b. *Net section* ($F_t = 0.5F_u = 29 \text{ ksi}$):

$$H = (7.5 - 2.0) \times 0.655 \times 29 = 104 \text{ kips} > 70 \qquad \text{OK}$$

4. *Check the bolt bearing stress:*

$$f_p = \frac{70}{4 \times 0.655 \times \frac{7}{8}} = 30.5 \text{ ksi} < F_p \qquad \text{OK} \quad \text{(AISCS, Sec. 1.5.1.5.3)}$$

5. *Bolts connecting tee to column flange* (neglect prying action in initial trial selection):

allowable tensile stress, A325 bolts, $F_t = 44 \text{ ksi}$ (AISCS, Table 1.5.2.1)

$$\text{trial number of bolts req'd} = \frac{70}{44 \times 0.601} = 2.65$$

Try four $\frac{7}{8}$-in.-diameter A325 bolts. Refer to Fig. 6.14.
Determine the prying force, by Eq. (6.27):

$$b = 2.0 - \frac{0.655}{2} - \frac{1}{16} = 1.61 \text{ in.}$$

$$F = \frac{70}{4} = 17.5 \text{ kips}$$

$$w = \frac{7.5}{2} = 3.75 \text{ in.}$$

$$a = \frac{11.265 - 4}{2} = 3.63 \text{ in.—but not to exceed } 2 \times 1.06 = 2.12 \text{ in.;}$$

$$\therefore \quad a = 2.12 \text{ in.}$$

Then, by Eq. (6.27),

$$Q = \left[\frac{100 \times 1.61 \times 0.875^2 - 18 \times 3.75 \times 1.06^2}{70 \times 2.12 \times 0.875^2 + 21 \times 3.75 \times 1.06^2} \right] \times 17.5 = 4.11 \text{ kips}$$

total bolt tension $= 17.5 + 4.1 = 21.6 \text{ kips} < 44 \times 0.601$ OK

6. *Recheck the bending stress in the tee flange by Eqs. (6.30) and (6.31):*

$$M_2 = 4.11 \times 2.12 = 8.71 \text{ kip-in.} \qquad \text{[Eq. (6.30)]}$$

$$M_1 = 17.5 \times 1.61 - 4.11 \times 2.12 = 19.46 \text{ kip-in.} \qquad \text{(governs)}$$

†Alternatively, shop welding could have been used.

$$f_b = \frac{6M}{bt_f^2} = \frac{6 \times 19.46}{3.75 \times 1.06^2} = 27.7 \text{ ksi} > 27$$

Here b is the length of tee per bolt. The flange is slightly overstressed—but in view of the known overconservatism in Eq. (6.27) for Q, 27.7 ksi should be acceptable.

B. *Shear connectors:*

1. *Angles to beam web*† (use $\frac{7}{8}$-in.-diameter A325 bolts in friction-type connection). The allowable double shear strength per bolt is

$$2F_v A_b = 2 \times 17.5 \times 0.601 = 21.04 \text{ kips}$$

$$\text{number of bolts req'd} = \frac{60}{21.04} = 2.85$$

Use three $\frac{7}{8}$-in.-diameter A325 bolts, angles to beam web. Use six $\frac{7}{8}$-in.-diameter A325 bolts, angles to column flange. (The single shear value of these is one-half double shear.)

2. *Required thickness of the shear connector angles* (try 9-in. length):

$$\text{req'd } t = \frac{V/2}{bF_v} = \frac{60/2}{9 \times 14.5} = 0.23 \text{ in.}$$

Here b is the length of the angles. Use two angles, $4 \times 3\frac{1}{2} \times \frac{5}{16} \times 9$ in. web angles.

C. *Design column web stiffeners, if needed:* Stiffeners will be designed at yield stress levels (see AISCS, Sec. 1.15.5) for the force P_{bf}, which is equal to the force H times $\frac{5}{3}$ for dead plus live load only and H times $\frac{4}{3}$ for the combination of loads that includes wind or earthquake. Hence

$$P_{bf} = \frac{4}{3} \times 70 = 93.3 \text{ kips}$$

In the compression zone, AISCS Formula (1.15-1) may be modified by adding $2t_f$ to $5k$, to allow for the additional spread of force at a $1:1$ slope through the flange of the tee section. t_b in the equation becomes the web thickness of the tee connector. The column web thickness is 0.34 in.

By the modified equation 1.15-1,

$$A_{st} > \frac{P_{bf} - F_{yc}t(t_b + 5k + 2t_f)}{F_{yst}} \ldots$$

$$= \frac{93.3 - 36 \times 0.34(0.655 + 5 \times 1.19 + 2 \times 1.06)}{36} = -0.37 \qquad \text{OK}$$

Additionally, check the column web clear depth in the compression zone, by AISCS Formula (1.15-2), for the column, for which

$$d_c = d - 2k = 9.98 - 2.38 = 7.60 \text{ in.}$$

$$\text{max. } d_c = \frac{4100 \times t^3\sqrt{F_{yc}}}{P_{bf}} = \frac{4100 \times 0.34^3\sqrt{36}}{93.3} = 10.4 \text{ in.} > 7.6 \qquad \text{OK}$$

No compression stiffeners are needed.

†Alternatively, shop welding could have been used.

In checking the need for a column web stiffener adjacent to the tension moment connector, AISCS Formula (1.15-3) is designed to apply to a plate welded directly to the column flange or to a bolted end plate. The four bolts in a tee connection would spread the load vertically along the column, and it could be assumed that for the 4-in. bolt spacing, plus one bolt diameter, a vertical length of about 9 in. of column web would come into play. The tensile web strength for this segment would be:

$$9 \times 0.34 \times 36 = 110.1 \text{ kips} > 93.3 \qquad \text{OK}$$

As far as direct tension is concerned, no stiffener in the tension zone appears to be needed. However, although not explicitly covered by the specification, the relatively small thickness of the W 10×49 column flange ($t = 0.56$ in.) suggests possible overstress in flange bending as a result of the outward pull of the bolts. The situation is similar to that sketched earlier in connection with the calculation for trial selection of tee flange thickness. (Rotated 90°, the tee web is now the column web.) Assuming that the column flange bending is fixed at the bolt line, that the vertical bolt gage is 4 in., and that a 4-in. vertical segment of column flange is effective, the bending moment due to one bolt would be $17.5 \times 1.0 = 17.5$ kip-in. and the local column flange bending stress would be

$$f_b = \frac{6 \times 17.5}{4 \times 0.56^2} = 83.7 \text{ ksi}$$

Obviously, to minimize local yielding, stiffeners should be introduced, similar to the upper plate c in Fig. 6.13(a). Note that, had the column flange been thicker than the flange of the tee connector, the check on local column flange bending could have been omitted.

Further study of moment-resisting connections designed for full continuity of frame action will be made in Chapter 8.

6.8 BOLTED END-PLATE CONNECTIONS

The bolted end-plate connection, illustrated in Fig. 6.15, has become increasingly popular. Shop-welded under controlled conditions and field-bolted, it can be designed for full end restraint and in many areas is the most economical type to use if continuity of frame action is utilized.

At the time of writing (late 1978), design practice had been patterned after the procedures described in Example 6.13 on the assumption that the upper and lower portions of the end plates are essentially similar in behavior to equivalent structural tees. However, the continuity of the full plate, its attachment to the beam web, and the fact that the design of the connection with structural tees is already based on overly conservative assumptions make

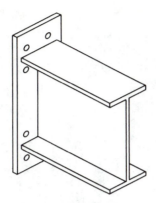

Fig. 6.15 Bolted end-plate connection.

use of this procedure grossly overconservative for end-plate design. The correctness of this statement has been demonstrated by recent extensive research sponsored in part by AISC. These studies indicate that end plates as currently designed may be anywhere from 40 to 60% thicker than is necessary.

The AISC is developing improved procedures for end plate design. It is recommended that reference be made to this topic in the 8th edition of the AISC *Manual* or to other special AISC publications for the latest information.

6.9 CONCLUDING REMARKS CONCERNING CONNECTIONS

As has been stated, connections can be extremely vital elements in a structure. Only a few standard types have been treated herein, and no textbook can really cover the subject adequately. The complexity and variety of connection details is infinite, and their design can only be partially covered by any specification. In special situations connection design requires the application of engineering judgment and experience to a much greater degree than does the design of simple beams and columns. Consider, for example, Fig. 6.16, showing a complex intersection of columns and diagonal elements, shop-welded and preassembled for proper fit up of high-strength bolted field connections. The weldment shown is part of a 70-ton assembly to support a 350-ft TV antenna on top of the north tower of the World Trade Center. The assembly provides the transition from the structural framing of the

Fig. 6.16 Shop assembly of welded and bolted connection for World Trade Center. (Courtesy Montague-Betts Company, Lynchburg, Virginia.)

building to the eight bearing plates for the antenna 12 ft above roof level. It was shop-assembled in an inverted position (as pictured) and all field connections were match-drilled.

PROBLEMS

6.1. Design a butt splice similar to that illustrated in Example 6.2, using $\frac{7}{8}$-in.-diameter A325 bolts in a friction-type connection, for A36 steel. Select plate sizes adequate to supply net section to transmit a load of 130 kips in tension and determine the number of bolts required.

6.2. Design a bracket connection similar to that of Example 6.3. Select two angles adequate for a tensile force of 130 kips, deducting for a single row of $\frac{7}{8}$-in.-diameter A490 bolts that connect the angles to the web of the WT. Design the connection between flanges of the tee and column using A490 friction-type bolts. Omit design for prying action. Use A36 steel.

6.3. Rework Example 6.4, changing the 160-kip horizontal force component to 180 kips and the 120-kip vertical force component to 135 kips.

6.4. Redesign the bracket connection, Problem 6.2, using welds made with E70 electrode. Follow the general pattern of Example 6.7, but include selection of angle sizes based on the tensile requirement for the gross section.

6.5. Assume in Example 6.4 that the horizontal component of force is transmitted to the pin plate and channel web in proportion to their respective thicknesses. Select a suitable fillet-weld size for the attachment of pin plate to channel web. Use E70 electrodes.

6.6. Design the eccentrically loaded bracket shown. Assume that there are two plates, one hidden from view on the far side of the column flanges, so that the 130-kip load is divided equally to the two plates. Use A490 $\frac{7}{8}$-in. friction-type high-strength bolts in single shear. Determine the required number. Refer to Example 6.8.

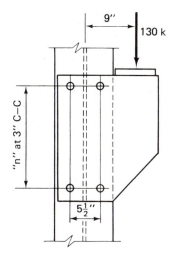

6.7. Redesign as welded the eccentrically loaded bracket of Problem 6.6 using the dimensions indicated, A36 steel, and E70 electrodes.

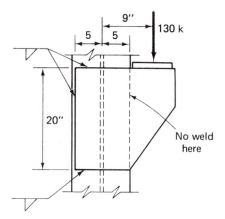

6.8. Same as Problem 6.7, except incline the 130-kip load outward and down at 60° from the horizontal.

6.9. Design an HS bolted seat and top angle beam-to-column connection for a reaction of 40 kips. The beam is a W 16 × 36 with $F_y = 50$ ksi. Use $\frac{7}{8}$-in. friction-type high-strength bolts.

6.10. Design a stiffened seat for a reaction of 65 kips using HS friction-type bolts, a W 16 × 36 beam, and $F_y = 50$ ksi.

6.11. Same as Problem 6.10, except do an all-welded design using E70 electrodes.

6.12. Following the pattern in the connection section of the AISCM, design a shop-welded and field-bolted moment connection for a W 24 × 84 beam framed to a W 12 × 65 column. The design moment is 320 kip-ft and the end reaction is 60 kips. All materials are ASTM A36 steel. Use A325 bearing-type bolts and E70 electrodes.

6.13. Using Example 6.13 as a guide, design a bolted "semirigid" beam-to-column connection, using top and bottom tees. A W 27 × 94 beam is connected to a W 14 × 90 column for a reaction of 70 kips and a bending moment of 120 kip-ft. All steel is type A36. Check for prying action to be included.

Note: A large number of additional problems may be developed simply by requiring a check verification of any of the tabular shear and/or moment values of connections described in the connection section of the AISCM.

7

PLATE GIRDERS

7.1 INTRODUCTION

Plate girders are built-up steel beams that require a section modulus greater than any available as a rolled beam. The most common form consists of two heavy flange plates between which is welded a relatively thin web plate. Girder depths range up to 20 ft or more and spans of several hundred feet are not unusual. At points of concentrated load or reaction the girder webs usually must be reinforced by *bearing stiffeners* to distribute the concentrated local forces into the web. *Intermediate* and/or *longitudinal stiffeners* may be added to serve in a quite different role—primarily that of increasing the buckling strength and thereby improving web effectiveness in resisting shear, moment, or combined stresses.

In recent years the economical maximum span of plate girders has greatly increased. This has been made possible in part by the reduction in required web thickness, which results from the use of the tension-field concept that permits utilization of the postbuckling strength of the girder web.

Plate girders are particularly favored for highway bridges; they permit unlimited vision and minimize clearance problems in traffic interchanges and complex multilevel overpasses. Plate girders are also frequently used in various types of buildings and industrial plants to support heavy loads. They are often used, for example, to provide a large space with no interfering columns on a lower floor of a high-rise building, as shown in Fig. 7.1.

Plate girders may be built up with bolts or rivets, as shown in Fig. 7.2(a) and (c), or welded, as shown in Fig. 7.2(b), (d), and (e). In situations where

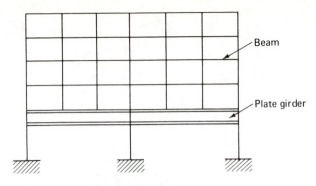

Fig. 7.1 Plate girders in building.

Fig. 7.2 Typical types of plate girders.

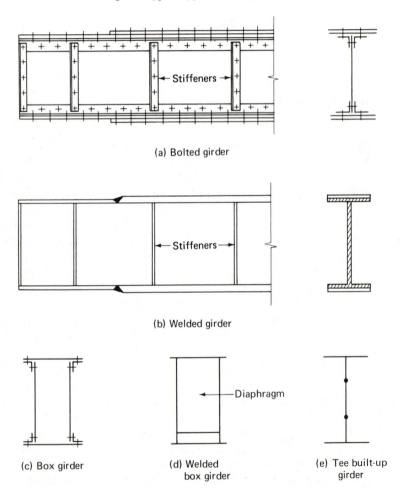

(a) Bolted girder

(b) Welded girder

(c) Box girder

(d) Welded box girder

(e) Tee built-up girder

lateral support of the compression flange cannot be provided, box girders [Fig. 7.2(c) and (d)] are especially recommended because of their superior effectiveness against lateral-torsional buckling and in resisting lateral loads. This results from their greater strength and stiffness in torsion and in bending about the weak axis. The design of box members in torsion as well as bending is treated in Chapter 10.

Figure 7.3 shows a long cantilever section of a plate girder being lifted into position for field splicing during the construction of an aircraft hangar.

Fig. 7.3 Plate girders supporting roof of United Airlines hangar in San Francisco, consisting of center span with 142-ft cantilevers. (Courtesy American Institute of Steel Construction.)

7.2 SELECTION OF GIRDER WEB PLATE

The selection of the web plate as shown in Fig. 7.4 involves the following steps:

1. Choose a web depth in relation to the span.
2. Choose the minimum thickness in terms of the permissible depth/thickness ratio.

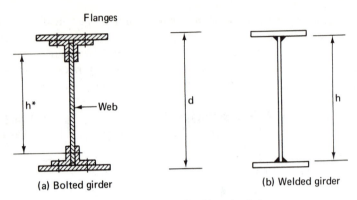

(a) Bolted girder   (b) Welded girder

*h as shown to be used as clear depth in
calculating web h/t ratios, but h = full web
depth in calculating web shear stress

Fig. 7.4 Typical plate girder.

1. Choice of Web Depth in Relation to Span

The depth of girders ranges from one-fourteenth to one-sixth of the length, depending on span and load requirements. The shallower girders are desirable if the service loads are light; deeper girders will be needed if the loads are heavy or if it is desired to keep deflections at a minimum. It may be desirable to make several preliminary designs with corresponding cost estimates to achieve an optimum depth.

2. Choose Web Thickness in Terms of Permissible Depth/Thickness Ratio

The AISCS for plate girders with intermediate stiffeners permits the girder web at allowable loads to go into the postbuckling range to develop tension-field action. After a stiffened thin web panel buckles in shear, it can continue to resist increasing load. The buckled web can resist diagonal tension (left half of Fig. 7.5) much as the diagonals (right half of Fig. 7.5) perform their function as tension members in a Pratt truss. The diagonal web tensions create compressive forces in the intermediate stiffeners, and these vertical stiffeners must be designed to meet this added requirement, thus acting as indicated in a manner analogous to the vertical members of the truss. After the initial buckling of a plate girder web, the girder stiffness decreases rapidly, and the girder deflection may reach a value appreciably greater than predicted by ordinary bending theory.

When the postbuckling strength of the web plate is utilized, the criterion for permissible depth/thickness ratio of web is still determined by buckling considerations, but these arise from the fact that under stress the curvature of a plate girder creates vertical compression in the web due to a downward

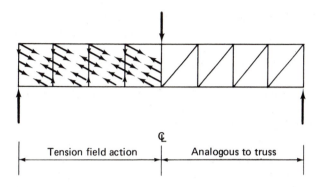

Fig. 7.5 Tension field action in plate girder web analogous to truss with tension diagonals.

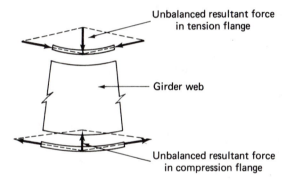

Fig. 7.6 Vertical web compression due to unbalanced force in flanges.

component of flange stress in a curved length of girder on the compression side and an upward component on the tension side, as shown in Fig. 7.6. The vertical buckling strength of the web plate must be sufficient to withstand this squeezing action, and this is taken care of if the depth/thickness ratio of the web meets the AISCS, Sec. 1.10.2, requirement that the ratio of the clear distance between flanges to the web thickness shall not exceed

$$\frac{h}{t} \leq \frac{14{,}000}{\sqrt{F_y(F_y + 16.5)}} \qquad \text{(see Table 7.1)} \qquad (7.1)$$

where

$h =$ clear distance between flanges (see the following note)
$t =$ web thickness

When the foregoing limitation is met, intermediate stiffener spacing as determined by the actual h/t and shear stress f_v is allowed to run as high as

3.0h. However, if intermediate stiffeners are held to an upper spacing limit of 1.5d, where d is the girder depth, the maximum permissible h/t ratio is greater:

$$\frac{h}{t} \leq \frac{2000}{\sqrt{F_y}} \qquad \text{(see Table 7.1)} \tag{7.2}$$

These upper limits on h/t are listed for various yield stress levels in Table 7.1, where it will be noted that the advantage of the 1.5d limit becomes more pronounced as the yield stress increases.

After selection of the web plate, the maximum shear stress $[f_{v(max)} = V_{max}/ht]$ should be determined. If $f_{v(max)}$ is less than F_v, by AISCS Eq. (1.10-1) (modified for indefinitely large stiffener spacing), and if, at the same time, h/t is less than 260, as well as being less than the limit provided by Eq. (7.1), no intermediate stiffeners are needed. Referring to AISCS, Formula (1.10-1), when a/h is very large, $k = 5.34$, and the allowable shear stress for no stiffeners is given by one of the following equations:

Table 7.1

Maximum Ratio of the Clear Distance Between Flanges to Web Thickness

F_y	36	42	46	50	55	60	65	90	100
Eq. (7.1)	322	282	261	243	223	207	192	143	130
Eq. (7.2)	333	309	295	283	270	258	248	211	200

When h/t is greater than $548/\sqrt{F_y}$,

$$F_v = \frac{83,150}{(h/t)^2} \tag{7.3}$$

When h/t is less than $548/\sqrt{F_y}$,

$$F_v = \frac{152\sqrt{F_y}}{h/t} \leq 0.4F_y \tag{7.4}$$

For example, what is the allowable shear stress for no stiffeners if $F_y = 36$ ksi and $h/t = 100$? In this case, $548/\sqrt{F_y} = 91.5$, and Eq. (7.3) applies, giving

$$F_v = \frac{83,150}{100^2} = 8.32 \text{ ksi}$$

This value, rounded off to 8.3, could also have been obtained from AISCS, Appendix A, Table 10-36, in the extreme right column for "a/h over 3." Equations (7.3) and (7.4) are useful in providing an accurate evaluation of F_v for values of h/t intermediate between those listed in Tables 10-36 and 10-50 and for steels with yield points other than 36 and 50 ksi.

Although not a controlling criterion, the designer at this point has the option of determining whether or not any reduction in allowable bending stress will be required as a result of tension-field action. By AISCS, Sec. 1.10.6, if h/t is less than $760/\sqrt{F_b}$, no reduction is required.

7.3 SELECTION OF GIRDER FLANGES

After the girder web is tentatively selected, the next step is to determine the sizes of the girder flanges. The steps in selecting girder flanges are:

1. Preliminary selection.
2. Check width/thickness ratios.
3. Determine the reduced allowable bending stress in flanges.
4. Select reduced-size flanges for use away from maximum moment, and determine the location of flange transitions.

1. Preliminary Flange Selection

As stated in Chapter 3, the required section modulus for a beam is

$$(S_x)_{\text{(req'd)}} = \frac{M_x}{F_b} \tag{7.5}$$

and the resisting moment supplied is

$$M_x = F_b(S_x)_{\text{(actual)}} \tag{7.6}$$

The bending strength of a plate girder equals the sum of the bending strengths of the girder web and flanges. An approximate evaluation of the girder's bending strength can be obtained as follows:

$$S_x = \frac{I_x}{d/2}$$

$$I_x = (I_x)_{\text{(web)}} + (I_x)_{\text{(flanges)}}$$

or, approximately

$$I_x \approx \frac{th^3}{12} + 2A_f\left(\frac{h}{2}\right)^2$$

where

$$t = \text{thickness of girder web}$$
$$h = \text{depth of girder web}$$
$$d = \text{depth of girder}$$
$$A_f = \text{area of one girder flange}$$

Then the section modulus of the girder for $h/d \approx 1$ becomes approximately

$$S_x = \frac{th^2}{6} + A_f h \tag{7.7}$$

Combining Eqs. (7.5), (7.6), and (7.7), the following expression for the required girder flange area is obtained:

$$A_f = \frac{M_x}{F_b h} - \frac{th}{6} \tag{7.8}$$

The first term on the right side of Eq. (7.8) represents the flange area that would be needed to resist the bending moment, M_x, without help from the web. The next term, $th/6$, is the *equivalent* flange area contributed by the girder web. Equation (7.8) provides a tentative trial selection that is subject to later verification by the moment-of-inertia method.

2. Trial Flange Plate Selection and b/t Check

After the tentative flange area is determined, a trial flange plate is selected. In order to prevent premature local flange buckling, maximum unsupported width/thickness ratios are imposed, as in column design, by AISCS, Sec. 1.9, as follows:

$$\frac{b}{t_f} < \frac{95}{\sqrt{F_y}} \qquad \text{for unstiffened plates}$$

$$\frac{b}{t_f} < \frac{238}{\sqrt{F_y}} \qquad \text{for stiffened plates}$$

(See AISCS, Appendix A, Table 6, for listed values.)

An unstiffened plate is one that is supported along one longitudinal edge and is free along the other, such as the half flange width of an I-shaped girder, or the outstanding width of a bearing stiffener. For the I-shaped girder, b is taken as equal to $b_f/2$, where b_f is the full flange width. The flange of a *box* girder is a stiffened plate, supported along both edges, for which b (in the absence of longitudinal stiffeners) is to be taken as the distance between the nearest lines of fasteners or welds. In light construction, using thin-walled girders, it is sometimes advantageous to exceed the above allowable b/t limits, which is permitted on the basis of reduced allowable stresses for which AISCS, Appendix C, provides information. These reductions are not to be confused with the mandatory reduction imposed by tension-field action, as discussed in the following section.

3. Determine Reduced Allowable Bending Stress in Flange

In tension-field design, the web is allowed to buckle, as previously discussed, and the stress on the compression side is no longer proportional to the distance from the neutral axis. As shown in Fig. 7.7, the stress at the

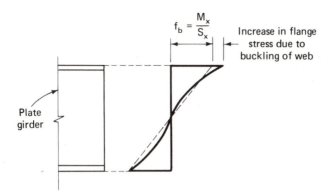

Fig. 7.7 Stress distribution after buckling of web.

extremity of the compression flange is slightly greater than given by the beam formula. To compensate for this in building design, a reduction in allowable flange stress is made according to Sec. 1.10.6 of the AISCS.

When the web depth/thickness ratio

$$\frac{h}{t} \geq \frac{760}{\sqrt{F_b}}$$

the maximum stress in the compression flange shall be

$$F'_b \leq F_b\left[1 - 0.0005\frac{A_w}{A_f}\left(\frac{h}{t} - \frac{760}{\sqrt{F_b}}\right)\right] \qquad \text{[AISCS Formula (1.10-5)]}$$

where

F_b = allowable bending stress given in the AISCS, Sec. 1.5.1
A_w = web area
A_f = compression flange area

After the trial flange selection is made, the moment of inertia and section modulus are calculated on the basis of actual dimensions and properties. The stress is then calculated and checked against the allowable value.

4. Select Reduced-Size Flanges for Use Away from Maximum Moment and Determine Location of Flange Area Transitions

For least weight in riveted and bolted girders the flange area should be in proportion to the bending moment. The size of the flange plates can be conveniently reduced in regions where the bending moments have decreased appreciably below the maximum values. However, the AISCS limits the cross-sectional area of cover plates of riveted girders to 70% of the total flange area. For light loadings it is probably reasonable to use a constant cover plate to run the full girder length. In the case of welded girders it is preferable to use a single flange plate, without flange angles or cover plates.

The flange plate thickness may be reduced at appropriate intervals, keeping in mind that such reductions should be made only if the saving in the cost of the flange material more than offsets the added expense of introducing the butt welds at thickness transition locations.

Cover plates in bolted girders (Fig. 7.8) can be cut off or flange plate

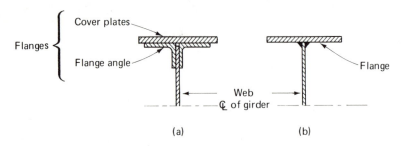

Fig. 7.8 Typical girder flanges.

thickness in welded girders reduced where the bending moments have dropped appreciably, as shown in Fig. 7.9. The resisting moments of the various girder sections where plates are cut off or reduced in thickness can be obtained by calculating the section modulus of the girder at each location [Eq. (7.5)]. One can locate the thickness transition point (or cutoff point) graphically by plotting horizontal lines that indicate the magnitude of the various resisting moments between transition points. The intersections of the moment diagram and the various horizontal lines, as shown in Fig. 7.9, determine the transition points. In the case of a uniform load the transitions can be determined mathematically from the equation for bending moment as follows:

$$\frac{x^2}{(l/2)^2} = \frac{M - M_r}{M} \quad \text{and} \quad \frac{M - M_r}{M} = \frac{A_c}{A}$$

where

$x =$ distance from the maximum moment in the girder to the theoretical thickness transition point (or cutoff point)
$l =$ span length of simply supported girder
$M =$ maximum moment in the girder due to uniform load
$M_r =$ resisting moment at theoretical thickness transition point
$A_c =$ area of plate to be cut off or difference of flange area between two flange plates at a transition point
$A =$ flange area plus web equivalent
$\quad = \dfrac{M}{hF_b}$ [see Eq. (7.8)]

Then the theoretical thickness transition point can be found:

$$x = \frac{l}{2}\sqrt{\frac{A_c}{A}} \tag{7.9}$$

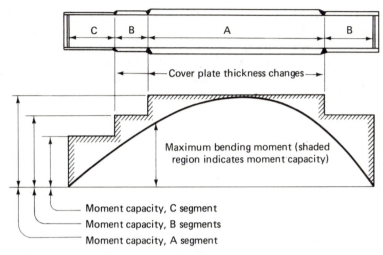

Fig. 7.9 Thickness change locations in plate girder flanges.

If cover plates are used in riveted, bolted, or welded girders, they should be extended beyond the theoretical transition points with enough fasteners or welds to develop the plate stress at the theoretical cutoff points. AISCS, Sec. 1.10.4, provides explicit information regarding extension requirements and details for flange development at the ends of cover plates.

In welded design with a single flange plate, transitions in section are best achieved by changing the flange plate thickness (and width, if desired) with the ends of the two flange plates joined by a full-penetration groove butt weld. The weld should have a concave contour that is smoothly tangent with the surface of the thinner plate.

If repeated load is a design requirement, the choice of details may be a critical factor. Referring to AISCS, Appendix B, the following comparisons are tabulated for loading condition 3, Table B-1, for the range 500,000 to 2,000,000 loading cycles:

	Illustrative case (AISCS)	Stress category (AISCS, Table B2)	Allowable stress range (ksi), load condition 3
Welded girder without stiffeners with no cover plates or splices	4	B	18
Welded groove weld transition (full penetration)	12	C	13
Welded girder with intermediate stiffeners	7	C	13
Welded girder with cover plates (stress at cutoff location)	5	E	8

It is readily seen that the use of cutoff welded cover plates may introduce a severe penalty under the conditions stipulated by the foregoing tabulation. If butt joints are used at transition points with full-penetration groove welds, the section may be changed where the reduced section stress would have a maximum stress *range* of no more than 13 ksi—in which case there might not be any penalty due to repeated load.

Flowchart 7.1 summarizes the logical steps to be followed in selecting the girder flange sizes.

7.4   INTERMEDIATE STIFFENERS

When the shear stress in the web is kept below the allowable value by AISCS Formula (1.10-1), intermediate stiffeners, if required, serve simply to improve the buckling resistance of the web plate and no tension-field action takes place. For greater shear stress, limited only by AISCS Formula (1.10-2), the shear strength is assumed to be the sum of the buckled web plate strength and the diagonal tension that is induced in the buckled web. The intermediate stiffeners then play the dual role of improving buckling resistance and acting as compression struts as in truss action.

The design of intermediate stiffeners, when such stiffeners are required, consists of

1. Plotting a maximum shear stress diagram.
2. Locating first stiffeners away from each end.
3. Locating the remaining intermediate stiffeners.
4. Selecting the size of the intermediate stiffeners.
5. Checking the maximum tensile stress in the web.

We consider these now in greater detail.

1. Plot Maximum Shear Stress Diagram

Stiffener spacing is a function of shear stress in the particular panel under consideration. It is also a function of the h/t thickness ratio, which has been established at the outset. With the aid of tables in AISCS, Appendix A, the plot of maximum shear stress along the girder permits rapid selection of intermediate stiffener spacing.

2. Locate First Stiffeners Away from Each End

In girder design the spacing between stiffeners at end panels and panels adjacent to those containing large holes shall be such that the web shear stress at end panels does not exceed the allowable shear stress given by AISCS

Flowchart 7.1

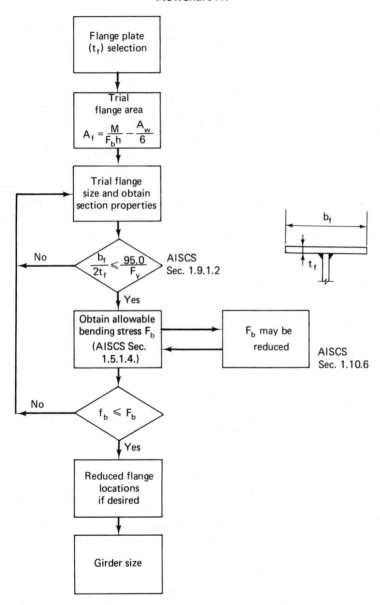

Flange plate
(t_f) selection

Trial
flange area

$$A_f = \frac{M}{F_b h} - \frac{A_w}{6}$$

Trial flange
size and obtain
section properties

No

$$\frac{b_f}{2t_f} \leq \frac{95.0}{\sqrt{F_y}}$$ AISCS
Sec. 1.9.1.2

Yes

Obtain allowable
bending stress F_b
(AISCS Sec.
1.5.1.4.)

F_b may be
reduced

AISCS
Sec. 1.10.6

No

$$f_b \leq F_b$$

Yes

Reduced flange
locations
if desired

Girder size

Formula (1.10-1) or as tabulated in AISCS, Appendix A, Table 10-36 and 10-50, for the 36-ksi and 50-ksi yield point steel, respectively. Thus these panels are designed without any benefit of tension-field action and are assumed to act as anchor panels for the adjacent tension fields.

3. Locate Remaining Intermediate Stiffeners

Referring to AISCS, Appendix A, Tables 10-36 and 10-50, one enters the table with the given h/t ratio and locates the column wherein the allowable shear stress as listed just exceeds the maximum shear stress in that particular column. The caption for the column then determines the allowable a/h, or stiffener spacing divided by clear depth of web. In these tables the allowable shear stress by AISCS Formula (1.10-1) is used for web design at stress levels below the buckling load, and Formula (1.10-2) tabulated in AISCS, Appendix A, Tables 11-36 and 11-50, for $F_y = 36$ ksi and $F_y = 50$ ksi, respectively, at stress levels above the buckling load. Above the buckling load the shear strength is the sum of the buckling strength plus that added by the tension field.

The need for intermediate stiffeners and (if needed) the determination of their proper location and size are complexly interrelated by the provisions of AISCS, Secs. 1.10.2 and 1.10.5.† The sequence of steps in making these determinations is summarized by Flowchart 7.2.

4. Select Size of Intermediate Stiffeners

According to Sec. 1.10.5.4 of the AISCS, a pair of intermediate stiffeners, or a single intermediate stiffener, shall be so selected as to satisfy

$$I_{st} \geq \left(\frac{h}{50}\right)^4$$

and

$$A_{st} \geq \frac{1 - C_v}{2}\left[\frac{a}{h} - \frac{(a/h)^2}{\sqrt{1 + (a/h)^2}}\right] YDht \qquad \text{[AISCS Formula (1.10-3)]}$$

where

C_v, a, h, and t are as defined previously
A_{st} = gross area of a single or pair of intermediate stiffeners, square inches
I_{st} = moment of inertia of a pair of intermediate stiffeners, or a single intermediate stiffener, with respect to an axis in the plane of the girder web
Y = (yield stress of web steel)/(yield stress of stiffener steel)
D = 1.0 for stiffeners furnished in pairs
  = 1.8 for single angle stiffeners
  = 2.4 for single plate stiffeners

---

†It is essential at this point that these provisions be studied independently in the AISCS in conjunction with Flowchart 7.2, as they will not be repeated in detail in this text.

Flowchart 7.2

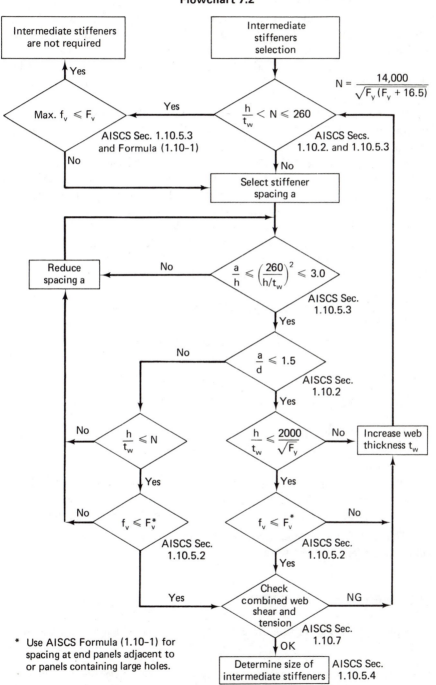

* Use AISCS Formula (1.10-1) for
  spacing at end panels adjacent to
  or panels containing large holes.

The moment of inertia limitation is intended to prevent local bending of the web between buckled panels. The area requirement is for the additional purpose of supplying adequate compression strut capacity during tension-field action.

Solutions of AISCS Formula (1.10-3) for required area are provided by the italicized listings in AISCS, Appendix A, Tables 11-36 and 11-50, for $F_y = 36$ ksi and $F_y = 50$ ksi, respectively.

It is noted that the intermediate stiffeners may be stopped short of the tension flange (AISCS, Sec. 1.10.5.4), provided that bearing is not needed to transmit a concentrated load or reaction. Near the tension flange the welds by which intermediate stiffeners are attached to the web shall be terminated not closer than four times the web thickness nor more than six times the web thickness from the near toe of the web to the flange weld. When single stiffeners are used, they shall be attached to the compression flange, if it consists of a rectangular plate, to resist any uplift tendency due to twist of the flange. Lateral bracing may be attached to the intermediate stiffeners, and these shall then be connected to the compression flange to transmit 1% of the total flange stress, unless the flange is composed only of angles.

5. Check Maximum Bending Tensile Stress in Girder Web

Plate girder webs, which depend upon tension-field action as considered in Formula (1.10-2), shall be so designed that bending tensile stress shall meet the requirement

$$f_{bx} \leq \left(0.825 - 0.375\frac{f_v}{F_v}\right)F_y \leq 0.6F_y \qquad \text{[AISCS Formula (1.10-7)]}$$

where

f_{bx} = bending tensile stress in webs due to moment in the plane of the girder web

f_v = computed average web shear stress (total shear divided by web area)

F_v = allowable shear stress according to AISCS Formula (1.10-2)

The use of this formula bypasses the more complex calculation of maximum direct tensile stress, which acts at an angle to the girder axis.

7.5 BEARING STIFFENERS

Bearing stiffeners serve three interrelated functions, which are illustrated in Fig. 7.10.

1. They distribute the transfer of local reactive forces to web shear, as illustrated in Fig. 7.10(a).

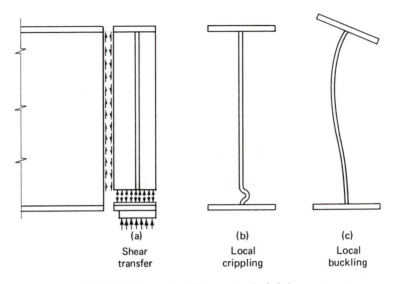

(a)	(b)	(c)
Shear	Local	Local
transfer	crippling	buckling

Fig. 7.10 Support conditions at end of girder.

2. They prevent local crippling in the web immediately adjacent to concentrated reactions or loads. This type of failure is illustrated in Fig. 7.10(b). If no bearing stiffeners are used, the local compressive stress in the web must be checked by AISCS Formula (1.10-8) for interior loads, or (1.10-9) for end loads or reactions. This topic has been treated previously in Section 3.9.

3. Finally, bearing stiffeners prevent a more general vertical buckling of the web, of the type illustrated in Fig. 7.10(c). In this connection the allowable average vertical stress components are specified by AISCS Formulas (1.10-10) and (1.10-11), the choice being determined by whether or not the top flange is restrained against rotation. In these formulas the vertical force is assumed to be distributed over a length of web equal to the girder depth or the length of the stiffened panel in which the load is placed—whichever dimension is the lesser.

Bearing stiffeners should have close contact with the flanges adjacent to application points of applied or reactive loads and should extend approximately to the edges of the flanges, as shown in Fig. 7.11. According to AISCS, Sec. 1.10.5.1, the stiffeners are to be designed as columns assuming the column section to comprise the pair of stiffeners and a centrally located strip of the web whose width is equal to not more than 25 times its thickness at interior stiffeners, or a width equal to not more than 12 times its thickness when the stiffeners are located at the end of the web. The effective length shall be taken as not less than three-fourths the length of the stiffeners in computing the slenderness ratio l/r. The stiffeners shall also be checked for local bearing pressure. Only that portion of the stiffeners outside the flange angle fillet or

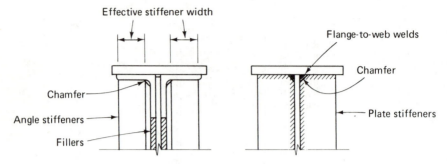

Fig. 7.11 Typical stiffeners.

the flange-to-web welds, as shown in Fig. 7.11, shall be considered effective in bearing and the bearing stress shall not exceed the allowable value of $0.90F_y$. (AISCS, Sec. 1.5.1.5.1).

Local detail is important in the transfer of a load concentration into bearing stiffeners. For example, if a very heavy column introduces a load or acts as a support, two pairs of stiffeners are desirable so as to introduce the column flange bearing stress directly into the stiffeners without local bending of the girder flanges, as shown in Fig. 7.12.

Fig. 7.12 Bearing stiffeners.

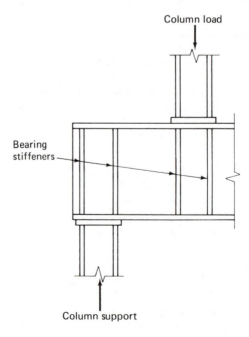

7.6 CONNECTIONS OF GIRDER ELEMENTS

1. Flange-to-Web Connection

Bolts or welds connecting flange to web, or cover plate to flange, shall be designed to resist the horizontal shear resulting from the bending forces on the girder, as shown in Fig. 7.13. The longitudinal spacing of bolts shall not exceed the maximum permissible provided in the AISCS, Sec. 1.18.2.3, for compression elements and in Sec. 1.18.3.1 for tension elements. The flange-to-web connection shall also transmit any direct loads that are applied unless bearing stiffeners are provided.

The flange-to-web connection shall be designed at all locations to transmit the shear force between flange and web due to the moment variation.

a. In the case of a bolted girder, as shown in Fig. 7.13(a), let

$P_1, P_2 =$ flange force due to moments M_1, M_2 at any two adjacent cross sections, respectively

$I =$ girder moment of inertia

$p =$ pitch of bolts

$R_x, R_y =$ force components of resultant shear force (R) on bolt

$V =$ shear force

$\bar{y} =$ distance between flange centroid and neutral axis of girder section

$w_y =$ vertical applied load per unit length

Then

$$R_x = P_2 - P_1 = \frac{M_2 - M_1}{I}\bar{y}A_f = \frac{Vp\bar{y}A_f}{I}$$

$$R_y = w_y p$$

$$R = \sqrt{R_x^2 + R_y^2}$$

b. In the case of the welded girder shown in Fig. 7.13(b), let

$q_x, q_y =$ components of stress resultant loads per unit length of fillet weld

$q_r =$ resultant shear between web and flange

Then

$$q_x = P_2 - P_1 = \frac{M_2 - M_1}{I}\bar{y}A_f = \frac{V\bar{y}A_f}{I}$$

$$q_r = \sqrt{q_x^2 + w_y^2}$$

2. Stiffener Connections

The welds, rivets, or bolts that attach bearing stiffeners to the girder web are simply designed to transmit the total reactive or applied load into the web. Similarly, the intermediate stiffeners in a tension-field panel must transfer

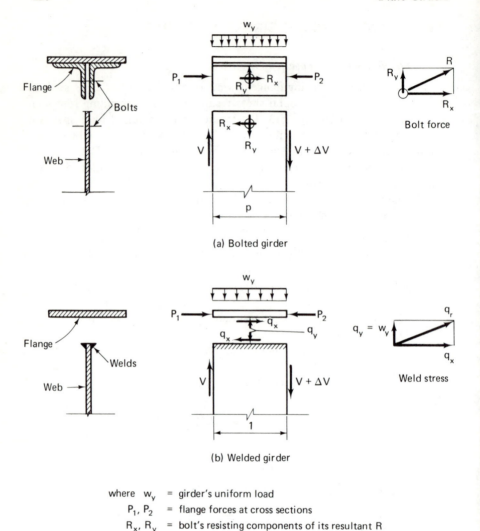

where w_y = girder's uniform load
P_1, P_2 = flange forces at cross sections
R_x, R_y = bolt's resisting components of its resultant R
q_x = stress intensity in horizontal direction
q_r = resultant shear between web and flange
p = bolt pitch

Fig. 7.13 Stress transmittance from girder flange to web.

the vertical component of the total tensile force into the stiffener. Such stiffeners, as required to meet the provisions of AISCS, Sec. 1.10.5.3, at shear stresses permitted by Formula (1.10-2) are to be connected to transmit at least

$$f_{vs} = h\left(\frac{F_y}{340}\right)^{3/2} \text{ kips per linear inch} \qquad \text{[AISCS Formula (1.10-4)]}$$

where

$$h = \text{as defined previously}$$
$$F_y = \text{yield stress of web steel}$$

According to Sec. 1.10.5.4 of the AISCS, the rivets connecting stiffeners to the girder web shall be spaced not more than 12 in. on center. For intermittent fillet welds connecting stiffeners to the girder web, the clear distance between welds shall not be more than 16 times the web thickness or more than 10 in.

3. Web Splices

Splices for girders offer no problem in welded girders, but are expensive and should be avoided whenever possible in riveted or bolted girders. The need for girder splices is dictated by erection requirements and shipping limitations. It may be desirable to locate the flange splice and web splice in different locations. The girder web primarily transmits shear force, and therefore web splices are most economically located at places where the shear force is small. Riveted or bolted splices should provide a net section area through the splice plates sufficient to resist the shear force and bending moment carried by the girder web at the splice location. Tests have shown the simple butt splice illustrated in Fig. 7.14 to be quite adequate.

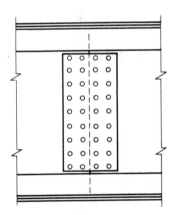

Fig. 7.14 Web splice for riveted or bolted girder.

In designing the web splice for moment, the web moment M_w, which is the portion of the total moment (M) carried by the girder web at the splice, can be obtained approximately from the proportion of the equivalent web area to flange, which has been introduced in Section 7.3; then

$$M_w = \frac{th/6}{(th/6) + A_f} M$$

After design shears and web moments are determined, the splice design for either bolted or welded connections follows standard procedures for connections, as covered in Chapter 6.

When fillers are needed in the splices, then fillers shall be designed according to Sec. 1.15.6 of the AISCS.

In welded girders, splices present no special problem; complete-penetration groove welds shall be provided to develop the full strength of the smaller spliced section.

7.7 ILLUSTRATIVE EXAMPLE

The information relative to plate girder design in the AISCM should be studied in conjunction with the following design example, which is similar to one provided in the AISCM, 7th edition.

This chapter may be further supplemented by reference to Example 9.5.

Example 7.1

Design a welded plate girder with a simple span of 56 ft to support a uniformly distributed load of 3 kips/ft (girder weight is included) and two concentrated loads

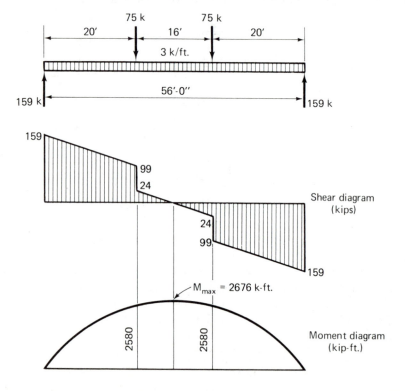

of 75 kips located 20 ft from each end. The compression flange is laterally supported only at points of concentrated load. Use A36 steel and the AISCS.

Solution

A. *Selection of girder web plate:*
1. *Select a girder with a depth of about one-eighth of the span length:*

$$\frac{l}{8} = \frac{56 \times 12}{8} = 84 \text{ in.}$$

Choose web depth $h = 80$ in.

2. *For no reduction in flange stress* (AISCS, Sec. 1.10.6):

$$\frac{h}{t} \leq \frac{760}{\sqrt{F_b}} = \frac{760}{\sqrt{22}} = 162$$

The corresponding thickness of web $= h/162 = \frac{80}{162} = 0.494$ in.

3. *Maximum ratio of clear distance between flanges to web thickness* (AISCS, Sec. 1.10.2):

$$\frac{h}{t} = \frac{14,000}{\sqrt{F_y(F_y + 16.5)}} = 322$$

The corresponding minimum thickness of web is

$$t = \frac{h}{322} = \frac{80}{322} = 0.238 \text{ in.}$$

Try web plate $\frac{1}{4} \times 80$; $A_w = 20$ in.²; $h/t = 80/\frac{1}{4} = 320$.

Note: Allowable flange stress will be reduced.

B. *Selection of girder flanges:*
1. *Preliminary flanges by flange-area method:* assume $F_b = 20$ ksi [reduction in flange stress is required; see item A(2)]:

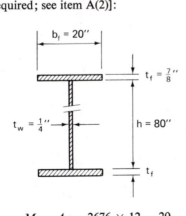

$$A_f = \frac{M}{F_b h} - \frac{A_w}{6} = \frac{2676 \times 12}{20 \times 80} - \frac{20}{6} = 16.8 \text{ in.}^2$$

Try

$$\text{plate } \tfrac{7}{8} \times 20 = 17.5 \text{ in.}^2 > 16.8 \text{ in.}^2 \quad \text{OK}$$

Section properties:
Moment of inertia:

flanges: $2 \times 20 \times \frac{7}{8} \times 40.438^2 = 57,233$
web: $\frac{1}{12} \times \frac{1}{4} \times 80^3 = 10,667$
$$I_x = 67,900 \text{ in.}^4$$

Section modulus:

$$S_x = \frac{67,900}{40.875} = 1661 \text{ in.}^3$$

Moment of inertia of flange plus $\frac{1}{6}$ web about weak y axis:

$$I_{oy} = \tfrac{1}{12} \times \tfrac{7}{8} \times 20^3 = 583 \text{ in.}^4$$

$$A_f + \frac{A_w}{6} = 17.5 + \tfrac{20}{6} = 20.83 \text{ in.}^2$$

$$r_T = \sqrt{\frac{583}{20.83}} = 5.3 \text{ in.}$$

Note: If the dead weight of the girder was an estimated part of the design load, it should be rechecked at this point on the basis of web and flange areas. Add 5% to allow for stiffeners and other details of welded girders.

2. *Check the width/thickness ratio* (AISCS, Sec. 1.9.1.2): the permissible flange width/thickness ratio is

$$\frac{b_f}{2t_f} < \frac{95}{\sqrt{F_y}} = 15.8 \qquad \text{(AISCS, Appendix A, Table 6)}$$

where

$$\frac{b_f}{2t_f} = \frac{20}{2 \times \frac{7}{8}} = 11.4 < 15.8 \qquad \textbf{OK}$$

3. *Determine the allowable bending stress in the flanges:*
   a. *For the 16-ft middle panel* [AISCS, Sec. 1.5.1.4.5(2)],

$$\sqrt{\frac{102 \times 10^3 C_b}{F_y}} = 53\sqrt{C_b} = 53 \qquad \text{where } C_b = 1$$

$$\frac{l}{r_T} = \frac{16 \times 12}{5.3} = 36.2 < 53$$

$$F_b = 0.6 F_y = 22 \text{ ksi}$$

The reduced allowable bending stress in the 16-ft panel is

$$F_b' = F_b\left[1 - 0.0005\frac{A_w}{A_f}\left(\frac{h}{t} - \frac{760}{\sqrt{F_b}}\right)\right] \qquad \text{[AISCS Formula (1.10-5)]}$$

$$= 22\left[1 - 0.0005\frac{20}{20.83}\left(320 - \frac{760}{\sqrt{22}}\right)\right] = 19.9 \text{ ksi}$$

The maximum actual bending stress is

$$f_b = \frac{M}{S_x} = \frac{2676 \times 12}{1661} = 19.3 \text{ ksi} < F_b' \qquad \text{OK}$$

b. *For the 20-ft end panels,*

$$C_b = 1.75 \times 1.05 \frac{M_1}{M_2} + 0.3 \left(\frac{M_1}{M_2}\right)^2$$

$$= 1.75$$

$$53\sqrt{1.75} = 70.4$$

where

$$\frac{M_1}{M_2} = 0$$

$$\frac{l}{r_T} = \frac{20 \times 12}{5.3} = 45.3 < 70.4$$

$$F_b = 22 \text{ ksi}$$

$$F_b' = 19.9 \text{ ksi}$$

The maximum actual bending stress is

$$f_b = \frac{M}{S_x} = \frac{2580 \times 12}{1661} = 18.6 \text{ ksi} < F_b' \qquad \text{OK}$$

Use for the web, one plate $\frac{1}{4} \times 80$; for the flanges, two plates $\frac{7}{8} \times 20$.

C. *Intermediate stiffeners:* Stiffeners are not required if $(h/t) < 260$ and $f_v < F_v$:

$$\frac{h}{t} = \frac{80}{\frac{1}{4}} = 320 > 260$$

Stiffeners are required:

$$f_v = \frac{V}{A_w} = \frac{159}{20} = 7.95 \text{ ksi}$$

1. *Locate the first stiffener from the end* (AISCS, Sec. 1.10.5.3). Try $a = 25$ in. ($a = 30$ in. was tried and found inadequate). Then

$$\frac{a}{h} = \frac{25}{80} = 0.3125 < 1.0$$

$$k = 4.0 + \frac{5.34}{(a/h)^2} = 58.68$$

$$C_v = \frac{45,000 \, k}{F_y(h/t)^2}$$

$$= 45,000 \times \frac{58.68}{36 \times 320^2} = 0.716 < 0.8 \qquad \text{OK}$$

Therefore,

$$F_v = \frac{F_y C_v}{2.89} = 36 \times \frac{0.716}{2.89} = 8.92 \text{ ksi} > f_v \qquad \text{OK}$$

2. *Locate remaining intermediate stiffeners* (AISCS, Secs. 1.10.5.2 and 1.10.5.3):

$$\frac{a}{h} \le \left(\frac{260}{h/t}\right)^2 = \left(\frac{260}{320}\right)^2 = 0.66 \qquad a \le 0.66 \times 80 = 52.8 \text{ in.}$$

a. For spacing between the end stiffener and the concentrated load, refer to AISCS, Appendix A, Table 11-36. Enter the table on the line for $h/t = 320$. Extrapolating beyond the column for $a/h = 0.6$, the allowable shear stress $F_v = 9.7$ ksi, for $a/h = 0.66$. Check the shear stress 25 in. from the end of the girder, at the first stiffener.

$$V = 159 - 3 \times 2.08 = 152.8 \text{ kips}$$

$$f_v = \frac{V}{A_w} = \frac{152.8}{20} = 7.64 \text{ ksi} < 9.7 \quad \text{OK}$$

Thus for the total distance between the first stiffener and the concentrated applied load, use five spaces at 43 in. each.

The following check on allowable shear stress is included to illustrate the computations that are involved:

$$\frac{a}{h} = \frac{43}{80} = 0.54$$

$$k = 4 + \frac{5.34}{(a/h)^2} = 22.3$$

$$C_v = \frac{45,000k}{F_y(h/t)^2} = 0.27 < 0.8$$

Since $C_v < 1.0$, the allowable shear stress is determined by AISCS, Sec. 1.10.5.2, Formula (1.10-2).

$$F_v = \frac{F_y}{2.89}\left[C_v + \frac{1 - C_v}{1.15\sqrt{1 + (a/h)^2}}\right]$$

$$= 10.3 \text{ ksi} \qquad \text{(See AISCS, Appendix A, Table 11-36)}$$

b. Distance in the midspan region between concentrated loads is a total of 16 ft. On the basis of previous calculations, a limit of 52.8 in. would be satisfactory. However, the only option is to use four spaces at 48 in., as three at 64 in. would exceed the limit appreciably. Check the maximum bending tensile stress in the girder web (Sec. 1.10.7):

$$F_{bx} = \left(0.825 - 0.375\frac{f_v}{F_v}\right)F_y \le 0.6F_y$$

$$= \left(0.825 - 0.375\frac{4.95}{10}\right)F_y = 0.64F_y$$

Use $F_{bx} = 0.6F_y = 22$ ksi > bending stress in web. Use a spacing of five at 43 in. between the end stiffener and the concentrated load, and a spacing of four at 48 in. in the center region.

3. *Select the size of the intermediate stiffeners* (AISCS, Sec. 1.10.5.4):
   Stiffener area:

$$A_{st} = \frac{1 - C_v}{2}\left[\frac{a}{h} - \frac{(a/h)^2}{\sqrt{1 + (a/h)^2}}\right] YDht \qquad \text{[AISCS Formula (1.10-3)]}$$

Or use AISCS, Table 11-36, for $h/t = 320$ and $a/h = 0.57$; then

$$A_{st} = 0.107ht = 0.107 \times 80 \times \tfrac{1}{4} = 2.14 \text{ in.}^2$$

Actual area required:

$$\frac{f_v}{F_v} A_{st} = \frac{7.64}{10.3} \times 2.14 = 1.59 \text{ in.}^2$$

Try

$$\text{two plates } \tfrac{1}{4} \times 4 = 2.0 \text{ in.}^2 > 1.59 \qquad \text{OK}$$

a. *Check the width/thickness ratio* (AISCS, Sec. 1.9.1.2):

$$\frac{b}{t} = \frac{4}{\tfrac{1}{4}} = 16 \approx 15.8 \qquad \text{OK}$$

b. *Check the moment of inertia* (AISCS, Sec. 1.10.5.4):

$$I_{st} \geq \left(\frac{h}{50}\right)^4 = \left(\frac{80}{50}\right)^4 = 6.55 \text{ in.}^4 \quad \text{required}$$

$$I_{st} = \tfrac{1}{12}0.25(2 \times 4 + 0.25)^3 = 11.7 \text{ in.}^4 \quad \text{furnished} > 6.55 \qquad \text{OK}$$

c. *Length required* (AISCS, Sec. 1.10.5.4):

$$h - 4t = 80 - 4 \times \tfrac{1}{4} = 79 \text{ in.}$$

Use for intermediate stiffeners two plates $\tfrac{1}{4} \times 4 \times 6$ ft 7 in. bearing on the compression flange of the girder.

D. *Bearing stiffeners* (AISCS, Sec. 1.10.5.1): Under concentrated loads and at the end of the girder, design for end reaction. Since bearing stiffeners must extend approximately to edges of flange plates, try two $\tfrac{1}{2} \times 8$ plates.

1. *Check the width/thickness ratio* (AISCS, Sec. 1.9.1.2):

$$\frac{b}{t} = \frac{8}{0.5} = 16 \approx 15.8 \qquad \text{OK}$$

2. *Check the web crippling* (AISCS, Sec. 1.10.10):
   a. Resulting from a distributed load of 3 kips/ft, flange assumed to be restrained from rotation:

$$f_a \leq \left[5.5 + \frac{4}{(a/h)^2}\right]\frac{10,000}{(h/t)^2}$$

where

$$f_a = \frac{P}{bt} = \frac{3}{12 \times \tfrac{1}{4}} = 1.0 \text{ ksi}$$

$$\left(5.5 + \frac{4}{0.54^2}\right)\frac{10,000}{320^2} = 1.88 \text{ ksi} > 1.0 \qquad \text{OK}$$

b. Resulting from a concentrated load of 159 kips at ends:

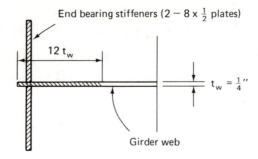

End bearing stiffeners (2 – 8 x $\frac{1}{2}$ plates)

12 t_w

$t_w = \frac{1}{4}''$

Girder web

$$I = \tfrac{1}{12} \times 0.5 \times 16.25^3 = 179 \text{ in.}^4$$

$$A_{\text{eff}} = 2 \times 0.5 \times 8 + 12 \times 0.25^2 = 8.75 \text{ in.}^2$$

$$r = \sqrt{\frac{I}{A}} = \sqrt{\frac{179}{8.75}} = 4.52 \text{ in.}$$

$$l = \tfrac{3}{4}h = \tfrac{3}{4} \times 80 = 60 \text{ in.}$$

$$\frac{l}{r} = \frac{60}{4.52} = 13.3$$

From AISCS, Appendix A, Table 3-36, for $Kl/r = 13.3$, then

$$F_a = 21.0 \text{ ksi}$$

$$f_a = \frac{P}{A_{\text{eff}}} = \frac{159}{8.75} = 18.2 \text{ ksi} < F_a \qquad \text{OK}$$

Use two plates $\frac{1}{2} \times 8 \times 6$ ft 8 in. as bearing stiffeners, with close fit against flanges that receive the concentrated loads. Use the same size of bearing stiffeners under concentrated loads.†

E. *Connections* (use an E70XX electrode):
  1. *Flange-to-web connection:*
     Horizontal shear:

$$q_x = \frac{V \bar{y} A_f}{I} = \frac{159 \times 40.44 \times 17.5}{67,900} = 1.66 \text{ kips/in.}$$

Vertical shear:

$$q_y = \tfrac{3}{12} = 0.25 \text{ kips/in.}$$

Resultant shear:

$$q_r = (q_x^2 + q_y^2)^{1/2}$$
$$= (1.66^2 + 0.25^2)^{1/2} = 1.68 \text{ kips/in.}$$

---

†For interior bearing, $25t$ may be used in determining the effective web area under concentrated loads at interior panels according to Sec. 1.10.5.1 of the AISCS.

Required weld size:

$$\frac{q_r/2}{0.707F_v} = \frac{1.68/2}{0.707 \times 21} = 0.057 \text{ in.}$$

Minimum weld size for a $\frac{7}{8}$-in. flange plate:

$$w_{min} = \tfrac{5}{16} \text{ in.} \qquad \text{(AISCS, Table 1.17.2A)}$$

Allowable strength for $\frac{5}{16}$-in. welds:

$$q_a = 0.707wF_v$$
$$= 0.707 \times \tfrac{5}{16} \times 21 = 4.64 \text{ kips/in.} > q_r \qquad \text{OK}$$

Minimum length of fillet welds (AISCS, Sec. 1.17.4):

$$l_{min} = 4w = 4 \times \tfrac{5}{16} = 1\tfrac{1}{4} \text{ in.}$$

Maximum spacing (AISCS, Sec. 1.18.3.1):

$$a_{max} = 24t \quad \text{or} \quad 12 \text{ in.} \qquad \text{whichever is smaller}$$
$$= 24 \times \tfrac{1}{4} = 6.0 \text{ in.}$$

Try $\frac{5}{16}$-in. welds, $1\frac{1}{2}$ in. long; then the required spacing is

$$a = \frac{2lq_a}{q_r} = \frac{2 \times 1.5 \times 4.64}{1.68} = 8.3 \text{ in.}$$

Use $\frac{5}{16} \times 1\frac{1}{2}$ in. welds, 6 in. on centers.

2. *Stiffener connections:*
   a. *Welds between web and intermediate stiffeners:*
      Transferred shear (AISCS, Sec. 1.10.5.4):

$$f_{vs} = h\left(\frac{F_y}{340}\right)^{3/2} = 0.034h = 0.034 \times 80 = 2.72 \text{ kips/in.}$$

Required weld size:

$$\frac{f_{vs}/2}{0.707F_v} = \frac{2.72/2}{0.707 \times 21} = 0.0915 \text{ in.}$$

Minimum weld size for $\frac{1}{4}$-in. plate is $\frac{1}{8}$ in. (AISCS, Sec. 1.17.2).
Allowable strength for $\frac{1}{4}$-in. welds:

$$q_a = 0.707 \times \tfrac{1}{4} \times 21 = 3.71 \text{ kips/in.}$$

Minimum length of fillet welds:

$$4w = 4 \times \tfrac{1}{4} = 1 \text{ in.}$$

Maximum spacing:

$$24t = 24 \times \tfrac{1}{4} = 6 \text{ in.} < 12 \text{ in.} \qquad \text{OK}$$

Try $\frac{1}{4}$-in. welds, $1\frac{1}{2}$ in. in length. Required spacing:

$$a = \frac{2lq_a}{f_{vs}} = \frac{2 \times 1.5 \times 3.71}{2.72} = 4.1 \text{ in.}$$

Use $\frac{1}{4} \times 1\frac{1}{2}$ in. welds, 4 in. on centers.

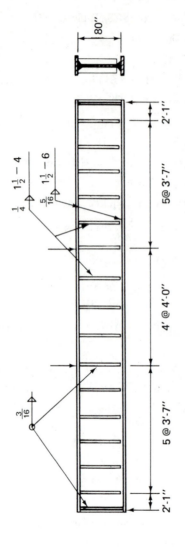

80"

$\frac{3}{16}$

$\frac{1}{4}$ $1\frac{1}{2} - 4$

$\frac{5}{16}$ $1\frac{1}{2} - 6$

2'-1" 5 @ 3'-7" 4' @ 4'-0" 5@ 3'-7" 2'-1"

Web: 1 plate $\frac{1}{4}$ × 80

Flanges: 2 plates $\frac{7}{8}$ × 20

Intermediate stiffeners : 2 plates $\frac{1}{4}$ × 4 × 6'-7"

Bearing stiffeners : 2 plates $\frac{1}{2}$ × 8 × 6'-8"

b. *Welds between bearing stiffeners and web:* Try minimum weld size, $\frac{3}{16}$-in. continuous welds both sides; then $\frac{3}{16}$-in. weld strength is

$$q_a = 0.707 \times \tfrac{3}{16} \times 21 = 2.78 \text{ kips/in.}$$

Total weld strength:

$$2q_a h = 2 \times 2.78 \times 80 = 444 \text{ kips} > 159 \text{ kips} \qquad \text{OK}$$

PROBLEMS†

7.1. For a welded girder cross section made up of single $\frac{7}{8} \times 22$ flange plates, top and bottom, connected by a $\frac{3}{8} \times 90$ web plate, determine:
   a. The approximate resisting moment based on the flange area method.
   b. The resisting moment by the moment of inertia method.
   c. The required stiffener spacing at a location where the shear is 300 kips.

7.2. The girder cross section shown is made up of an MC 18×42.7 channel attached to the top flange of a W 33×118 section by means of friction-type A325 high-strength bolts of $\frac{3}{4}$-in. diameter. What bolt pitch is required at a location where the shear on the girder is 220 kips?

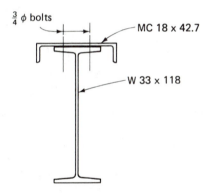

$\frac{3}{4} \phi$ bolts

MC 18 x 42.7

W 33 x 118

7.3. On the basis of the structural details illustrated on the next page, using E60 welding electrodes, determine the permissible end reaction R (same as end shear V), in each of the following five different ways:
   a. On the basis of the fillet weld size and weld spacing at the end of the girder, attaching web to flange plates.
   b. On the basis of the local contact compressive bearing stress at the bottom ends of the bearing stiffeners.
   c. On the basis of the size and arrangement of bearing stiffeners. (Assume that welds between bearing stiffeners and web are adequate.)
   d. On the basis of the first space (40 in.) from the bearing stiffener to the first intermediate stiffener.

†Assume use of A36 steel in all problems.

e. On the basis of the second space (60 in.) between the first and second intermediate stiffeners. (The shear used in the design of this space is assumed to be 20 kips less than the end reaction; that is, $V = R - 20$.)

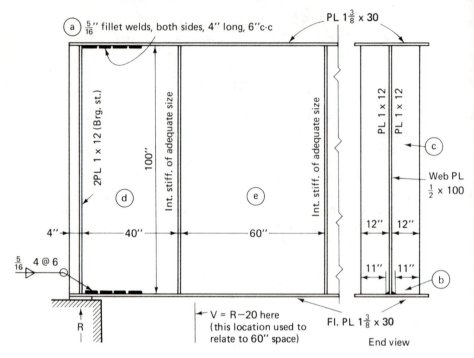

7.4. Design an all-welded plate girder for the following conditions. Include a general design drawing and detailed sketches of such portions as are judged needed to provide complete information. Span, 136 ft. Simple supports. A36 steel. Uniform live load of 3 kips/ft. Concentrated load of 760 kips, 26 ft from left support. Assume adequate lateral support. Web plates are to be as thin as possible, in commercially available thicknesses, to carry the maximum shear. Web plate thickness may be varied, if desired, to suit the high shear area at one end.

Note: This is a rather lengthy assignment and is suitable as a term project. For classroom use it is suggested that groups of three or four students each be assigned arbitrary web plate depths, between 110 and 160 in., varying in increments of 10 in. A plot of girder weight versus web plate depth can then be made as an exercise in the search for the most economical proportion.

The following is a summary† of steps that may be followed in carrying out this assignment:

1. Select the web plate:
   a. Choose a clear depth in relation to the span.

†Parenthetical references are AISCS sections.

   b. Choose the thickness. (1.10.2)
   c. Check for shear. (1.10.5.2)
   d. Check the maximum tension stress in the web. (1.10.7)

2. Flange plates:
   a. Make the preliminary selection by the flange-area method.
   b. Determine the reduced allowable stress. (1.10.6)
   c. Check the stresses due to bending by Mc/I.
   d. Select reduced size flanges for use away from the maximum moment and determine the location of the flange transitions. (Check the flange width/ thickness ratios.) (1.9.2.2)

3. Intermediate stiffeners:
   a. Locate the first stiffener away from each end. (1.10.5.3)
   b. Locate the remaining intermediate stiffeners. (1.10.5.2 and 1.10.5.3)
   c. Select the size of the intermediate stiffeners. (Check the area and I requirements.) (1.10.5.4)

4. Bearing stiffeners:
   a. Design for maximum reaction. (1.10.5.1)
   b. (For assigned problem, assume other bearing stiffeners to be identical.)

5. Design welds:
   a. Stress transfer for intermediate stiffeners. (1.10.5.1)
   b. Bearing stiffeners.
   c. Web-to-flange shear transfer.

6. Weight takeoff for complete girder.

8

CONTINUOUS BEAMS AND FRAMES

8.1 INTRODUCTION

In the design of a continuous structure, attention is turned from the individual member to the complete structure and the interrelated behavior of all its members. Continuous beams and frames in steel take advantage of the inherent continuity of all-welded construction. Weight reduction will result from the partial equalization of positive and negative moments at midspans and over supports, respectively. Deflections are reduced, the need for special bracing to resist lateral forces may be eliminated, and greater resistance to ultimate collapse due to earthquake or other shock loads is achieved.

The design of a continuous beam or frame may be based on elastic analyses and the allowable stress method, as covered by Part 1 of AISCS, or, alternatively, it may be based on plastic design for ultimate strength, as treated in Part 2.

In the elastic range of behavior, continuous beams and frames are statically indeterminate. The well-known method of moment distribution will be reviewed and used for analysis in some of the examples. In other cases tabular information is available in the AISCM for a limited number of span arrangements. If one is not familiar with methods for the analysis of continuous structures, the inclusion of these examples and problems is optional. Examples will omit design of details of types covered in earlier chapters, with primary attention given to member-size selection.

Coverage herein is limited to continuous beams and single-story frames. Multistory building design would involve complexities going beyond desired basic limits. Emphasis is on design examples that are specially selected to illustrate important concepts.

8.2 MOMENT-DISTRIBUTION ANALYSIS: A SUMMARY

The following condensed treatment of the *moment-distribution* method of analysis for continuous beams and frames is provided as a preliminary to the presentation of allowable stress design examples. The moment-distribution method provides a simple procedure to determine, through a converging iterative process, the bending moments at supports or joints in a continuous beam or frame. Once these "end moments" are known, the shears, reactions, and moments can be determined at any location by statics. The method is readily applied to problems involving sidesway. The procedure for frames under lateral force will be demonstrated when sidesway is induced by a lack of symmetry of applied vertical load.

The procedure is summarized as follows:

1. Assume the members to be locally locked against rotation at all support points and joints.
2. Enter the "fixed-end moments," using available tabular information or deriving them by basic beam analysis procedures. An important feature is the systematic entry of numerical data in a standardized tabular format. In entering the fixed-end moments, a special "rotational" sign convention is usually used; that is, fixed-end moments are positive if applied to the end of a member in a clockwise sense, negative if applied counterclockwise.
3. "Unlock" the joints. If any given joint is "unlocked," the ends of the members entering the joint, along with the joint, will rotate until equilibrium is established. This process introduces equilibrating moments in each member entering the joint in proportion to their relative rotational stiffness. As the joint rotates, there is a "carryover" of moment to the far end of each member if that member is either temporarily locked against rotation (as may be assumed) at the far end or if the far end is actually assumed to be permanently fixed against rotation. For a member of uniform cross section it is readily shown that the *carryover factor* is equal to $\frac{1}{2}$.
4. At those joints assumed to be temporarily locked against rotation, the moments introduced by carryover from the opposite ends will usually introduce a new unbalance of overall joint moment. In this case steps (3) and (4) must be repeated, in sequence, until the residual unbalanced moments at all joints are inconsequential to the calculation accuracy desired for design purposes.

The foregoing procedure is best reviewed in connection with numerical examples, such as those in Examples 8.2 and 8.5.

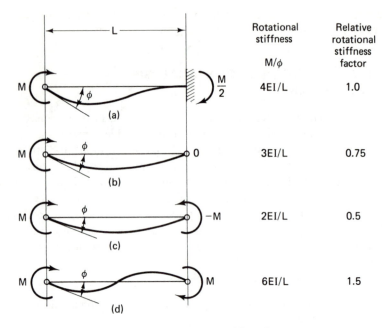

	Rotational stiffness M/ϕ	Relative rotational stiffness factor
(a)	$4EI/L$	1.0
(b)	$3EI/L$	0.75
(c)	$2EI/L$	0.5
(d)	$6EI/L$	1.5

Fig. 8.1 Moment rotational stiffness factors.

The basic concepts of rotational stiffness and carryover are illustrated in Fig. 8.1(a), where a positive (clockwise) moment M_A is introduced at A, causing (by "carryover") a positive moment at fixed end B of magnitude $M_A/2$. The rotational stiffness at A, or M_A/ϕ_A, is equal to $4EI/L$. If member AB is located and loaded in a frame in such a way that the relation between ϕ_A and ϕ_B is known in advance, due to symmetry, antisymmetry, or because the end at B is hinged, modified rotational stiffness values permit a reduction in the required computation because no carryover needs to be made in the particular span involved. The modified stiffness values for these special cases are tabulated in Fig. 8.1(b)–(d), along with their relative values normalized with respect to the basic case shown in Fig. 8.1(a). The use of these modifications to expedite the analysis is illustrated in Examples 8.2 and 8.5.

8.3 ALLOWABLE-STRESS DESIGN OF CONTINUOUS BEAMS

In the allowable-stress procedure, elastic continuous beam analyses are made to determine maximum positive and negative bending moments for various critical positions of live load. If the chosen W beam section meets the compact section requirements of AISCS, Sec. 1.5.1.4.1, the beam may be "proportioned for $\frac{9}{10}$ of the negative moments produced by gravity loading which are maximum at points of support, provided that, for such members, the maximum positive moment shall be increased by $\frac{1}{10}$ of the average negative moments." This adjustment in moments is justified by the fact that when yielding starts at the supports, due to negative moment, the positive moments increase at a more rapid rate, and by the time the failure load is reached the positive and negative moments are more or less equalized.

In Example 8.1 advantage is taken of the availability of tables of moments for continuous beams of two, three, or four spans of equal length in the AISCM. The final result should be identical with that based on continuous beam analyses, as in examples that follow. In Example 8.6, plastic design will be used for the same problem.

Example 8.1

Select a W beam of A36 steel for a continuous beam of three equal spans of 30 ft each. A dead load of 1 kip/ft (including weight of beam) and live load of 2 kips/ft are assumed to be uniform. Use the allowable-stress procedure. Simple supports are assumed at extreme ends. Continuous lateral support will be provided by the floor system.

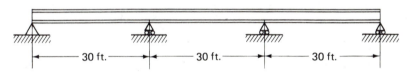

Solution

Calculate moments due to dead load (AISCM Tables).
 Positive moment in end span, $0.4l$ from end:

$$M = 0.080wl^2 = 0.080 \times 1.0 \times 30^2 \times 12 = 864 \text{ kip-in.}$$

Negative moment at supports:

$$M = -0.100wl^2 = -0.10 \times 1.0 \times 30^2 \times 12 = -1080 \text{ kip-in.}$$

Calculate moments due to live load (AISCM Tables) with only end spans loaded for maximum positive moment in end span.
 Maximum positive moment $0.45l$ from end:

$$M = 0.1013wl^2 = 0.1013 \times 2.0 \times 30^2 \times 12 = 2188 \text{ kip-in.}$$

Negative moment for same load condition (not maximum):

$$M = -0.050wl^2 = -0.05 \times 2.0 \times 30^2 \times 12 = -1080 \text{ kip-in.}$$

Maximum negative moment (AISCM Tables); one end span unloaded:

$$M = -0.1167wl^2 = -0.1167 \times 2.0 \times 30^2 \times 12 = -2521 \text{ kip-in.}$$

Make beam selection for $\frac{9}{10}$ maximum negative moment or for maximum positive moment increased by $\frac{1}{10}$ of the average negative moments (AISCS, Sec. 1.5.1.4.1).

Negative design moment:

$$M_{des} = 0.9(1080 + 2521) = 3241 \text{ kip-in.}$$

Positive design moment (maximum dead and live load moments are conservatively assumed to be at the same location):

$$M_{des} = (864 + 2188) + 0.1\left(\frac{0 + 2160}{2}\right) = 3160 \text{ kip-in.}$$

Negative moment controls beam selection. Section modulus required:

$$S_x = \frac{3241}{24} = 135 \text{ in.}^3$$

Referring to the allowable-stress beam selection tables, AISCM, a W 21 × 68 is selected. $S_x = 140$ in.^{3}. No limits are indicated on F'_y, hence thickness ratios are satisfactory for the assumed allowable stress of $0.66F_y$ (AISCS, Sec. 1.5.1.4.1). Check the shear stress. The maximum is at the end of the loaded middle span when one end span is unloaded (refer to AISCM Tables):

$$V_{max} = 0.6 \times 1.0 \times 30 + 0.617 \times 2.0 \times 30 = 55.0 \text{ kips}$$

$$f_v = \frac{55.0}{21.13 \times 0.43} = 6.06 \text{ ksi} < 14.5 \qquad \text{OK}$$

Note: If the shear stress is quite large, it might be desirable to check the direct stress in the web, adjacent to the flange fillet, at the interior support. The procedure in Example 10.3, Eq. (10.13), could be used, but the check is not required by the AISCS and the stress is usually not critical.

The complete design would also involve a check on the need for bearing stiffeners over the supports, their design if needed, along with bearing plates and other framing details. As mentioned earlier, examples in this chapter will be concerned primarily with member-size selection, as the design of the details has been covered previously. Whether or not splices are needed depends on availability and shipping restrictions—if needed, complete continuity can be provided by a full-penetration butt weld, and the splice or splices should be located in a region of low bending moment.

In a second example the same overall length and loads will be designated, but the interior supports will be moved toward the ends by 1 ft 6 in., making the center span 33 ft. Since the AISCM tabular information is no longer applicable, the required analyses will be made by moment distribution, as summarized in Section 8.2. Only two analyses will be required, (1) load on one end span, and (2) load on the center span; all load combinations thereby follow by superposing results of these solutions and by taking advantage of the overall symmetry of the structure.

Example 8.2

Same as Example 8.1, except for new interior support locations, as shown in the sketch. Determine the bending moments at the interior supports with the left span loaded with 1 kip/ft.

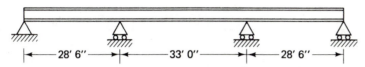

Solution

Since the moment of inertia is constant over all three spans, only relative values of the beam stiffness factors (I/L) are needed. Choose any convenient value for I, 1000, for example. Since the end spans are hinged at their extremities, the stiffness modification factor of $\frac{3}{4}$, as explained in connection with Fig. 8.1(b), will be introduced. The distribution factors, in fractions or decimal parts of unity, are entered at each support, and the F.E.M. (fixed-end moments) are calculated and entered. The F.E.M. at the hinged supports are balanced and half of the balancing moments are carried over to the interior supports, thenceforth all carryover and balancing of moments is confined to the center span. The tabulation of the foregoing operations follows:

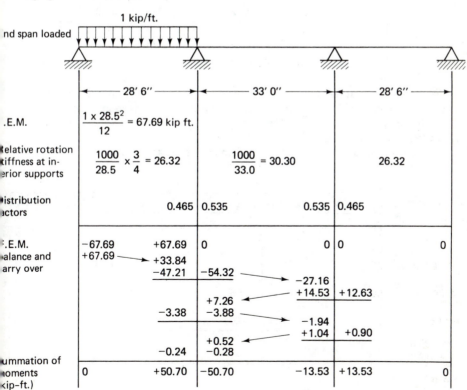

We next apply uniform load of 1 kip/ft to the center span, in which case complete symmetry of load and structure permit use of modified stiffness factors for all spans. Final moments are obtained in one balancing operation with no carryover, since there are no unbalanced moments.

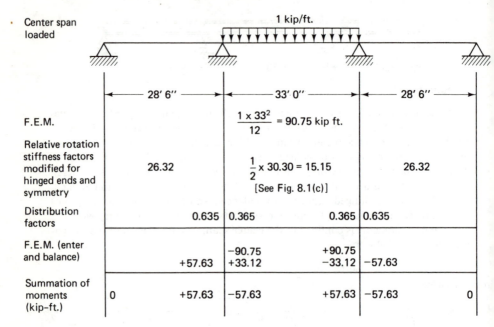

Center span loaded			1 kip/ft.		
		← 28′ 6″ →	← 33′ 0″ →	← 28′ 6″ →	
F.E.M.			$\frac{1 \times 33^2}{12}$ = 90.75 kip ft.		
Relative rotation stiffness factors modified for hinged ends and symmetry		26.32	$\frac{1}{2}$ × 30.30 = 15.15 [See Fig. 8.1(c)]	26.32	
Distribution factors		0.635	0.365 · · · 0.365	0.635	
F.E.M. (enter and balance)		+57.63	−90.75 / +33.12 · · · +90.75 / −33.12 · −57.63		
Summation of moments (kip-ft.)	0	+57.63	−57.63 · · · +57.63 · −57.63		0

The moments at the supports due to the dead load alone will now be determined as the summation of (1) the left end span loaded, (2) the center span loaded, and (3) the right end span loaded, obtained by reversing, end for end, the results for the left end span loaded.

Moments at supports due to the dead load alone are as follows:

Loaded span			1 kip/ft.		
Left end span	0	+50.70	−50.70 · · · −13.53	+13.53	0
Center span	0	+57.63	−57.63 · · · +57.63	−57.63	0
Right end span	0	−13.53	+13.53 · · · +50.70	−50.70	0
Summation, dead load moments (kip-ft.)	0	+94.80	−94.80 · · · +94.80	−94.80	0

The maximum moment that determines the required beam size will be the greatest of the following, after modification as allowed by AISCS, Sec. 1.5.1.4.1, for compact sections.

1. Maximum positive moment in end span; put the live load on both end spans.
2. Maximum positive moment in center span; put the live load on the center span only.
3. Maximum negative moment at support; put the live load on one end span and the center span.

The moments at the supports due to the live load of 2 kips/ft are obtained by direct 2:1 proportion from the previous solutions.

1. *Maximum positive moment in the end span:*

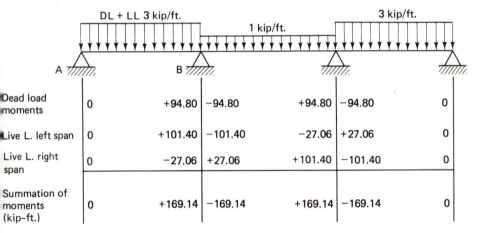

	A	B				
Dead load moments	0	+94.80	−94.80	+94.80	−94.80	0
Live L. left span	0	+101.40	−101.40	−27.06	+27.06	0
Live L. right span	0	−27.06	+27.06	+101.40	−101.40	0
Summation of moments (kip–ft.)	0	+169.14	−169.14	+169.14	−169.14	0

The maximum positive moment in the end span will be at the location of zero shear. Determine the left reaction. Consider the left span as isolated and take moments about the right end:

$$R_A \times 28.5 + 169.14 - 3 \times 28.5 \times 14.25 = 0$$

Solving for R_A,

$$R_A = 36.82 \text{ kips}$$

Let x be the distance from A to the location of zero shear:

$$R_A - 3x = 0 \qquad x = 12.27 \text{ ft}$$

Maximum moment:

$$M_{\max} = 36.82 \times 12.27 - 3 \times \frac{12.27^2}{2} = 226.0 \text{ kip-ft}$$

Maximum moment increased for design (AISCS, Sec. 1.5.1.4.1):

$$M_{\text{des}} = 226 + \frac{0 + 169.14}{2 \times 10} = 234.5 \text{ kip-ft}$$

2. *Maximum positive moment in the center span:*

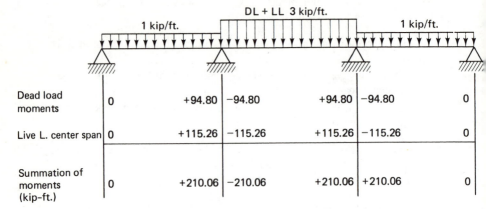

Dead load moments	0	+94.80	−94.80	+94.80	−94.80	0
Live L. center span	0	+115.26	−115.26	+115.26	−115.26	0
Summation of moments (kip-ft.)	0	+210.06	−210.06	+210.06	+210.06	0

The maximum positive moment is at the center of the center span and is equal to the simple beam moment: $wL^2/8$ less 210.06 kip-ft.

Maximum moment:

$$M_{max} = \frac{3 \times 33^2}{8} - 210.06 = 198.32 \text{ kip-ft}$$

Maximum moment increased for design (AISCS, Sec. 1.5.1.4.1):

$$M_{des} = 198.32 + \frac{210.06}{10} = 219.3 \text{ kip-ft}$$

3. *Maximum negative moment at the interior support:*

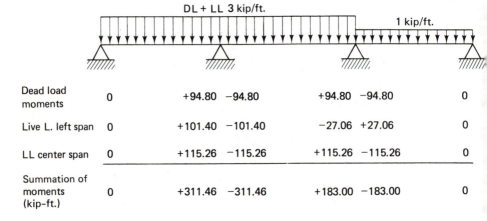

Dead load moments	0	+94.80	−94.80	+94.80	−94.80	0
Live L. left span	0	+101.40	−101.40	−27.06	+27.06	0
LL center span	0	+115.26	−115.26	+115.26	−115.26	0
Summation of moments (kip-ft.)	0	+311.46	−311.46	+183.00	−183.00	0

The maximum negative moment reduced for design (AISCS, Sec. 1.5.1.4.1) is

$$M_{des} = 0.9 \times 311.46 = 280.31 \text{ kip-ft} \qquad \text{(controls selection)}$$

The required section modulus is

$$S = \frac{280.31 \times 12}{24} = 140.16 \text{ in.}^3$$

Use a W 21 × 68, $S = 140$ in.^{3}.

The design of the support details is omitted. It may be noted that the movement of supports from the positions in Example 8.1 has had very little effect on the maximum design moment and that the same beam selection is made as before.

As an alternative to a continuous beam, an articulated series of beams with alternate spans cantilevering over the interior supports may be considered. Such a structure has the advantage of being statically determinate and is unaffected by settlement of supports, a contributing factor in its widespread selection for highway overpass bridges. Hinges are logically placed near locations that would have zero moment if the same structure were continuous. Example 8.3 provides a comparison with Examples 8.1 and 8.2 by using the same overall length and load conditions.

Example 8.3

Here we are dealing with a three-span structure using articulated beams, 90 ft 0 in. overall. Use A36 steel, 2 kip/ft live load, 1 kip/ft dead load.

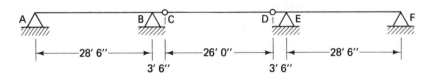

Solution

For maximum positive moment in span *AB*, put the live load on segment *AB* and take moments about support *B*, noting that the dead load on *CD* causes a concentrated reactive load of 13 kips at *C*:

$$R_A \times 28.5 + 13 \times 3.5 + \frac{1 \times 3.5^2}{2} - \frac{3 \times 28.5^2}{2} = 0$$

Solving,

$$R_A = 40.94 \text{ kips}$$

Let x be the distance from *A* to the location of zero shear.

$$40.94 - 3x = 0$$

$$x = 13.65 \text{ ft}$$

$$M_{\max} = 40.94 \times 13.65 - 3 \times \frac{13.65^2}{2} = 279.3 \text{ kip-ft}$$

The maximum positive moment in span CD,

$$M_{max} = \frac{3 \times 26^2}{8} = 253.5 \text{ kip-ft}$$

The maximum negative moment is at B or E. The entire length may be loaded with live load. Take moments about B of cantilever segment BC, noting the reactive force of 39 kips at C.

$$M_{max} = 39 \times 3.5 + \frac{3 \times 3.5^2}{2} = 154.9 \text{ kip-ft} \qquad \text{(does not control)}$$

Beam size required for end spans AC and DF:

$$S = \frac{279.3 \times 12}{24} = 139.7 \text{ in.}^3$$

Use a W 21 $\times$ 68, $S = 140$ in.^{3}.

Beam size required for center cantilever span CD:

$$S = \frac{253.5 \times 12}{24} = 126.8 \text{ in.}^3$$

Use a W 21 $\times$ 62, $S = 131$ in.^{3}.

The beam size is the same in the end spans as in Examples 8.1 and 8.2. For the center segment there is a weight saving of 6×26 lb—not great, and the cost saving would be offset by the cost of the two hinged connections. However, this in turn might be counterbalanced by the probable need to introduce one or two all-welded field splices with full continuity in the case of the designs of Examples 8.1 and 8.2.

As a fourth design alternative, consider the use of three simply supported beams, 30 ft each in span.

$$M_{max} = \frac{3 \times 30^2}{8} = 337.5 \text{ kip-ft}$$

$$S = \frac{337.5 \times 12}{24} = 168.8 \text{ in.}^3$$

Use W 24 $\times$ 76 beams, $S = 176$ in.^{3}.

Thus an appreciable saving in weight is achieved in comparison with simple beam design, by use of either continuity or cantilevered articulation. Moreover, especially in the case of the continuous beam, the maximum deflections will be appreciably reduced as compared with simple beam design.

8.4 ALLOWABLE-STRESS DESIGN OF CONTINUOUS FRAMES

Attention will be given primarily to one-story frames such as are widely used in shops, supermarkets, and industrial plants. Similar design considerations are required in the case of multistory frames, but that specialized field is beyond the scope of this introduction to basic design. As in the case of continuous beams, the moment-distribution method will be applied and many of the design examples will also be treated subsequently by the alternative plastic-design procedure.

Special attention will now be given to columns that are simply roof support elements and are lacking in continuity with the main frame. Initially, to illustrate the problem, we consider the simplest possible situation of the column that must depend on its neighboring framed column or columns for lateral support at the top.

Referring to Fig. 8.2(a), column 1 is shown in a buckled configuration caused by vertical critical load P_1. At the instant of buckling the external

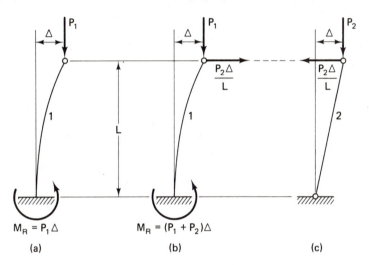

Fig. 8.2 Lateral support requirements for a hinged-end column supported by a neighboring column.

bending moment $P_1\Delta$ at the base is in exact equilibrium with the internal resisting moment developed at the base by the member. Now consider column 2, in Fig. 8.2(c), hinged at both top and base. If it is not supported laterally at the top, it will collapse as a mechanism under any load. The force required to provide lateral support, as shown at the top, must keep it in static equilibrium and is determined by taking moments about the lower hinge. It is equal to $P_2\Delta$.

Now assume that column 1 is attached at the top by a horizontal member to column 2 as shown in Fig. 8.2(b) and (c). In this case it must provide the lateral restraining force necessary to maintain equilibrium, and the required resisting moment at the base, as shown in Fig. 8.2(b), must be equal to

$$M_R = (P_1 + P_2)\Delta$$

This moment is exactly equal to the resisting moment that would be needed if column 1 were itself loaded with both loads P_1 and P_2. This fact leads to the following design procedure, which has been used in practice when hinged-end columns must be supported by their neighbors:

1. Design the hinged-end columns for their own applied loads with an effective length factor $K = 1$.

2. Framed or fixed-base columns that are connected at the top to hinged-end columns, so as to have the same lateral deflection at the top, shall be designed for their own column load plus the load, or share of the load, apportioned from otherwise unsupported hinged-end columns.

The foregoing procedure is quite accurate in the elastic or buckling range of behavior. In the inelastic range it is conservative but is recommended as a safe design procedure.

The following numerical Example 8.4 will further explain the discussion of Fig. 8.2.

Example 8.4

The structure shown is supported by the wall and roof bracing system to prevent lateral movement at the top of the columns out of the plane of the sketch. Select column sizes for the 100-kip center load using A36 steel W shapes.

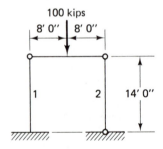

Solution

Select the size for column 1: Each column carries an actual load of 50 kips. Column 1 provides lateral support at the top for column 2, so it will be designed for 100 kips. Referring to Table 4.1, for buckling as a cantilever column, as shown, $K = 2.0$ in theory, but 2.1 is recommended for design to allow for some base rotation. Normal to the sketch, with lateral support at top, $K = 0.7$ in theory, with 0.8 recommended for design.

Try a W 8 × 28:

$$r_x = 3.45 \text{ in.} \qquad r_y = 1.62 \text{ in.}$$

$$A = 8.25 \text{ in.}^2$$

$$\frac{KL}{r_x} = \frac{2.1 \times 14 \times 12}{3.45} = 102.3 \quad \text{(governs)}$$

$$\frac{KL}{r_y} = \frac{0.8 \times 14 \times 12}{1.62} = 83.0$$

allowable load $= 12.68 \times 8.25 = 104.6 \text{ kips} > 100$ OK
(AISCS, Appendix A, Table 3-36)

Select the size for column 2: K = 1 for buckling in either direction. Design for an actual load of 50 kips.
Try a W 8 $\times$ 21:

$$A = 6.16 \text{ in.}^2 \qquad r_y = 1.26 \text{ in.}$$

$$\frac{KL}{r_y} = \frac{1.0 \times 14 \times 12}{1.26} = 133.3 \qquad \text{(governs)}$$

allowable load $= 8.40 \times 6.16 = 51.7 \text{ kips} > 50$ OK
(AISCS, Appendix A, Table 3-36)

In the following example a three-span single-story frame will be designed for an assumed dead load of 1 kip/ft and a live load of 2 kips/ft. As in the case of Example 8.2, a complete study would include a check on the maximum positive moments in the center and end spans and the negative moment at an interior support. The negative moment does, in fact, control the design, and the computations for maximum positive moment will, therefore, be omitted. For a maximum negative moment at an interior support, the live load is placed on the adjacent end and center spans, as in Example 8.2. This loading, being unsymmetrical, will cause sidesway, and the solution by moment distribution can be obtained in two steps: (1) introduce an imaginary constraint at the top end of the frame to temporarily eliminate sidesway, and (2) make a second solution by applying any horizontal load, with the constraint removed, and determine the horizontal force that is accounted for by the analysis. The complete solution is determined by superposing a factored proportion of the second solution with the first such that the force in the imaginary constraint is reduced to zero. Application of the procedure follows in Example 8.5.

Example 8.5

Select column and beam member sizes for the frame shown for a live load of 2 kips/ft and an assumed dead load of 1 kip/ft. Use A36 steel.

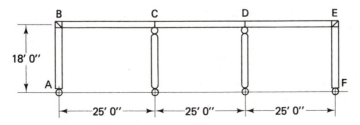

Solution

Introduce an imaginary horizontal constraint at location E to prevent sidesway. To simplify the tabulation of numerical work, the columns may be folded upward and outward. Assuming a uniform moment of inertia throughout, the relative stiffnesses with respect to support rotation are inversely proportional to span length. Modification factors apply only to the end spans as per Fig. 8.1(b). For the 25-ft spans, the dead load fixed-end moment is $1 \times 25^2/12 = 52.1$ kip-ft.

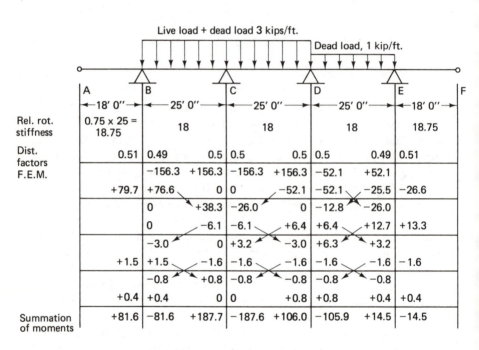

Live load + dead load 3 kips/ft. — Dead load, 1 kip/ft.

Spans: A $\leftarrow$ 18' 0" $\rightarrow$ B $\leftarrow$ 25' 0" $\rightarrow$ C $\leftarrow$ 25' 0" $\rightarrow$ D $\leftarrow$ 25' 0" $\rightarrow$ E $\leftarrow$ 18' 0" $\rightarrow$ F

	A	B		C		D		E
Rel. rot. stiffness	0.75 x 25 = 18.75	18		18		18		18.75
Dist. factors	0.51	0.49	0.5	0.5	0.5	0.5	0.49	0.51
F.E.M.		−156.3	+156.3	−156.3	+156.3	−52.1	+52.1	
	+79.7	+76.6	0	0	−52.1	−52.1	−25.5	−26.6
		0	+38.3	−26.0	0	−12.8	−26.0	
	0	0	−6.1	−6.1	+6.4	+6.4	+12.7	+13.3
		−3.0	0	+3.2	−3.0	+6.3	+3.2	
	+1.5	+1.5	−1.6	−1.6	−1.6	−1.6	−1.6	−1.6
		−0.8	+0.8	−0.8	−0.8	−0.8	−0.8	
	+0.4	+0.4	0	0	+0.8	+0.8	+0.4	+0.4
Summation of moments	+81.6	−81.6	+187.7	−187.6	+106.0	−105.9	+14.5	−14.5

The force in the imaginary constraint is now determined on the basis of the amount required to balance the horizontal shears at A and F. Taking moments about B and E of the end column segments, AB and EF, the horizontal base shears are determined, and these in turn determine the amount of force in the imaginary restraint at E that is required for overall horizontal equilibrium of the complete structure. Results are as follows:

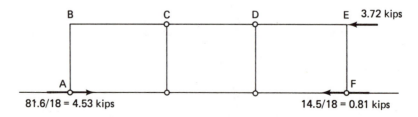

E 3.72 kips

81.6/18 = 4.53 kips   14.5/18 = 0.81 kips

We now analyze the same frame with the imaginary constraint removed and with a horizontal force of arbitrary magnitude applied at B but with the joints at B and E assumed not to rotate. If a force of 1 kip is chosen, the horizontal shear at A and F will be 0.5 kip each, and the bending moments at B and E at the tops of each column will be 9 kip-ft. The joints at B and E are now allowed to rotate and the usual moment distribution carryover and balance process is followed. Advantage may be taken of antisymmetry, as shown in Fig. 8.1(d), in which case the rotational stiffnesses at C and D in the central span are modified by a 1.5 factor and balancing moments are not carried over in CD.

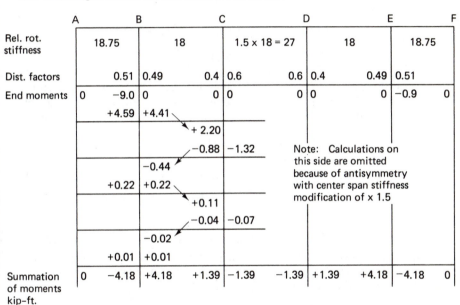

	A	B		C		D		E		F
Rel. rot. stiffness		18.75		18		1.5 × 18 = 27		18		18.75	
Dist. factors		0.51	0.49	0.4	0.6	0.6	0.4	0.49	0.51	
End moments	0	−9.0	0	0	0	0	0	0	−0.9	0
		+4.59	+4.41	+ 2.20						
				−0.88	−1.32					
			−0.44							
		+0.22	+0.22	+0.11						
				−0.04	−0.07					
			−0.02							
		+0.01	+0.01							
Summation of moments kip-ft.	0	−4.18	+4.18	+1.39	−1.39	−1.39	+1.39	+4.18	−4.18	0

Note: Calculations on this side are omitted because of antisymmetry with center span stiffness modification of x 1.5

The moments of 4.18 kip-ft at the top of columns AB and EF now account for base shears of $4.18/18 = 0.232$ kip at A and F, or a total equivalent force at B of 0.464 kip. In order to eliminate the imaginary restraining force of 3.72 kips in the first part of the solution, it is therefore necessary to multiply the results of the second solution by a factor of $3.72/0.464 = 8.02$ and superpose the factored results of both solutions.

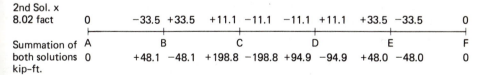

2nd Sol. x 8.02 fact	0	−33.5	+33.5	+11.1	−11.1	−11.1	+11.1	+33.5	−33.5	0
	A	B		C		D		E		F
Summation of both solutions kip-ft.	0	+48.1	−48.1	+198.8	−198.8	+94.9	−94.9	+48.0	−48.0	0

The summed moments at B and E are opposite in sense and equal in magnitude, within roundoff error limits, thus providing a partial check on overall equilibrium of the structure.

The maximum design moment at the interior support, assuming the member meets the compact section requirements of AISCS, Sec. 1.5.1.4.1.

$$M_{des} = 0.9 \times 198.8 = 178.9 \text{ kip-ft}$$

$$S_{req} = \frac{178.9 \times 12}{24} = 89.5 \text{ in.}^3$$

Try a W 21 × 50:

$$S = 94.5 \text{ in.}^3$$

From AISCM:

$$A = 14.7 \text{ in.}^2 \qquad b_f = 6.53 \text{ in.} \qquad r_x = 8.18 \text{ in.}$$

$$\frac{b_f}{2t_f} = 6.1 \qquad \frac{d}{t} = 54.8 \qquad \frac{d}{A_f} = 5.96$$

Check the column segment of the frame, member *AB*. To provide sidesway lateral support for hinged-end interior columns, assume *AB* to support half of full load on all three spans:

$$P = 3 \times \frac{75}{2} = 112.5 \text{ kips}$$

$$f_a = \frac{112.5}{14.7} = 7.65 \text{ ksi}$$

$$\frac{f_a}{F_y} = \frac{7.65}{36} = 0.213 > 0.16$$

This indicates use of AISCS Formula (1.5-4b), to check d/t, but, since the actual value of f_a is about $\frac{1}{3}$ of this, it may be used in AISCS Formula (1.5-4a). The added bending stiffness is the real requirement in supporting the interior columns, and d/t will be determined by the real value of f_a.

$$\frac{b}{2t_f} < \frac{65}{\sqrt{F_y}} = 10.8 > 6.1 \qquad \text{OK}$$

$$\frac{d}{t} < \frac{640}{\sqrt{F_y}}\left(1 - 3.74\frac{f_a}{F_y}\right) = 78.4 > 54.8 \qquad \text{OK}$$

Lateral supports to compression flange of column or beam components of main continuous frame should not exceed $76.0 b_f/\sqrt{F_y} = 76 \times 6.53/\sqrt{36} = 82.7$ in. or $20,000/[(d/A_f)F_y] = 20,000/[5.96 \times 36] = 93.2$ in.

Check the column segments for combined stress (AISCS, Sec. 1.6.1, and this text, Section 5.3). Determine K ($I_c = I_g = I$).

at the top,

$$\frac{I_c}{L_c} = \frac{I}{18} \qquad \text{and} \qquad \frac{I_g}{L_g} = \frac{I}{25}$$

By Eq. (4.5),

$$G_A = \frac{25}{18} = 1.39$$

At the base hinge (see Section 4.3),

$$G_B = 10$$

By Fig. 4.4(b),

$K = 1.95$

$$K\frac{L}{r_x} = 1.95 \times 18 \times \frac{12}{8.18} = 51.5$$

$F_a = 18.21$ ksi   (AISCS, Appendix A, Fig. 3-36)

$$\frac{f_a}{F_a} = \frac{7.65}{18.21} = 0.420 > 0.15 \quad \text{[AISCS Formulas (1.6.1a) and (1.6.1b) apply]}$$

$C_m = 0.85$

$F'_e = 56.32$ ksi   (AISCS, Appendix A, Table 9)

$$f_{bx} = \frac{48.1 \times 12}{94.5} = 6.11 \text{ ksi}$$

Check by AISCS Formula (1.6-1a):

$$\frac{7.65}{18.21} + \frac{0.85 \times 6.11}{(1 - 7.65/56.32)24} = 0.67 < 1 \quad \text{OK}$$

Check by AISCS Formula (1.6-1b):

$$\frac{7.65}{21.6} + \frac{6.11}{24} = 0.61 < 1 \quad \text{OK}$$

The load condition used for the foregoing slightly underestimates the maximum moment in the columns at B or E, but the design check shows much strength to spare. A change to a smaller column size would complicate the analysis and increase the moments in the main spans. In a complete design all critical load conditions should be included.

Design the interior hinged-end columns supporting frame at C and D. The critical load condition for maximum reactions is the same as for maximum moment. Calculate the maximum reaction:

$$R_{\max} = 3 \times 25 + \frac{198.8 - 48.1}{25} + \frac{198.8 - 94.9}{25} = 85.2 \text{ kips}$$

Try W 8 × 31:

$$A = 9.12 \text{ in.}^2 \qquad r_y = 2.01 \text{ in.} \qquad K = 1$$

$$\frac{KL}{r_y} = 18 \times \frac{12}{2.01} = 107.5$$

$F_a = 12.0$ ksi   (AISCS Appendix A, Table 3-36)
allowable load $= 12.0 \times 9.12 = 109$ kips   OK

This result is overly strong, but no smaller rolled W shape will do.

Attention is now turned to plastic design, with application to a number of the same design problems that have been treated elastically in the foregoing sections.

8.5 INTRODUCTION TO PLASTIC DESIGN

Before engaging in the study of plastic design for continuous beams and frames, one should review earlier material that is especially relevant. The whole concept of plastic design is dependent on the great ductility of structural steel and on its unique yield property as exhibited by the plastic portion of the stress–strain diagram shown in Fig. 1.2, the discussion of which in Section 1.3 should be reread at this time. In allowable stress (elastic) design, the maximum estimated stress is kept less than the specification yield point by a *factor of safety*. In plastic design, the ultimate strength of the structure is estimated and the allowable load is equal to a fractional part of the estimated ultimate strength. When the ultimate strength is reached, various localized parts of the structure will have yielded to a varying degree and finally, when collapse is imminent, the structure is said to have become a plastic mechanism. In a true mechanism (such as one of the interior columns of Example 8.5) there is no strength whatever unless lateral support is provided. In a plastic mechanism there is resistance to collapse, but deformation proceeds with no increase in load, just as it does in the simple tension test in the plastic range of Fig. 1.2.

Members used in plastic design are primarily those wide-flange W shapes that not only meet the requirements for a compact section of AISCS, Sec. 1.5.1.4.1, but are more compact to a degree that will permit bending deformation well into the plastic range without loss of compressive strength due to plastic buckling. This added compactness gives the member a quality known as plastic *rotation capacity*. The bending behavior of the W shape in the plastic range was discussed in detail in Section 3.3 in connection with the plastic design of simple beams that are statically determinant in both the elastic and plastic stages of their load-deflection history. While there can be no objection to the plastic design of simple beams, neither is there any particular advantage. In the simple beam the relative distribution of moment throughout the beam is the same in the elastic and plastic stages. Not so for the continuous beam or frame! After yielding is initiated at the location of maximum elastic-range moment, the relative distribution of moment along the continuous beam or frame starts to change. The total load on the structure continues to increase, although the deflection for a given increment of load becomes progressively greater as the plastic mechanism or collapse load is approached. It is this redistribution of moment that provides the potential for greater economy when plastic design is used.

The specification requirements for plastic design are covered in Part 2 of the AISCS. In the plastic design of a continuous beam or frame, the required member size is chosen on the basis of the required plastic modulus Z (see Section 3.3) that will provide the plastic moment needed to make the maximum strength of the structure greater than the desired "working" load multiplied by a load factor of "1.7 times the given live load and dead load or 1.3 times these loads acting in conjunction with 1.3 times any specified wind or earthquake forces" (AISCS, Sec. 2.1). Plastic design should not be used if the number of cycles of repeated load (AISCS, Appendix B) reaches a number requiring a limit on the maximum stress range. Plastic design ensures an optimum of structural integrity, or "toughness," against failure—a desirable feature in earthquake- or blast-resistant structures. Moreover, especially in the case of continuous beams, the structural analysis for plastic design is simplified because it is *statically determinate*. A complete and authoritative commentary on plastic design is provided in American Society of Civil Engineers Manual 41, "Plastic Design in Steel."

Before considering the alternative plastic design procedure applied to the conditions of Example 8.1, the ultimate strength moment–load relationships for an end span and for an interior span of a continuous beam will be examined. Assume that the M versus ϕ curve for the W section shown in Fig. 3.9 may be replaced by a two-straight-line approximation, as shown in Fig. 8.3.

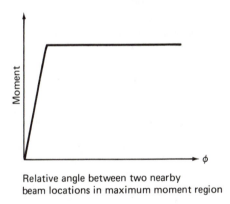

Relative angle between two nearby
beam locations in maximum moment region

Fig. 8.3 Assumed beam behavior in plastic design analysis.

For the end span shown in Fig. 8.4, elastic analysis shows that M_p would first be reached at the support where the maximum moment occurs. According to Fig. 8.3, this moment would not change with increasing load, and the maximum load condition would be reached when a positive moment of M_p was also reached, thus making this beam segment a plastic mechanism. It can be shown that the maximum positive moment is $0.414L$ from the simply supported end and that, for an end span,

$$M_p = 0.086wL^2 \tag{8.1}$$

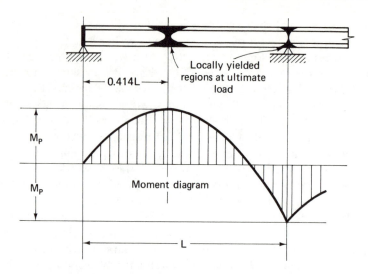

Fig. 8.4 End-span moments at ultimate plastic load.

For an interior span, M_p would be reached simultaneously at both ends, and after additional increase in load would also be reached at the center. Since the total range of moment is the same as the center moment in a simple beam, it is readily seen (Fig. 8.5) that

$$M_p = \frac{wL^2}{16} \tag{8.2}$$

Fig. 8.5 Interior span moments at ultimate plastic load.

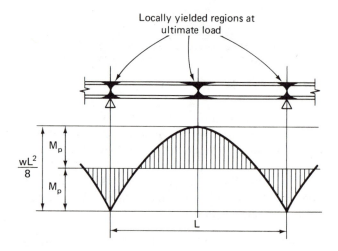

Example 8.6

Same loads and spans as for Example, 8.1, but design by Part 2 of AISCS. The total factored load on any one of the spans will be

$$1.7 \times 3.0 = 5.10 \text{ kips/ft}$$

Solution

Since the interior and end spans are of the same length, the beam selection will be determined by Eq. (8.1), which requires a greater M_p than Eq. (8.2). The required M_p, by Eq. (8.1), is

$$M_p = 0.086 \times 5.10 \times 30^2 \times 12 = 4.737 \text{ kip-in.}$$

The required plastic modulus is

$$Z = \frac{4737}{36} = 131.6 \text{ in.}^3$$

Referring to the AISCM, try a W 24 $\times$ 55, with $Z = 134$ in.^{3}. Check the width/thickness ratios for requirements of AISCS, Sec. 2.7:

$$\frac{b}{2t_f} = 6.96 < 8.5 \quad \text{OK}$$

$$\frac{d}{t_w} = 59.5 < 68.7 \quad \text{OK}$$

Check the shear capacity. In the end span, adjacent to the interior support, maximum shear occurs equal to

$$V_{\max} = \frac{wL}{2} + \frac{M_p}{L} = (0.50 + 0.086)wL = 0.586wL$$

$$= 0.586 \times 5.1 \times 30 = 89.7 \text{ kips}$$

Shear capacity (AISCS, Formula 2.5-1):

$$F_v = 0.55 \times 36 = 19.8 \text{ ksi}$$
$$V = 19.8 \times 0.395 \times 23.57 = 184.3 \text{ kips} > 89.7 \quad \text{OK}$$

Example 8.7

Same as Example 8.6, but with supports relocated to reduce end spans to an optimum condition that would make the required M_p for the end spans the same as that for the center span.

Solution

Let $x =$ length of end span. Then $90 - 2x =$ length of center span. By Eqs. (8.1) and (8.2),

$$0.086wx^2 = \tfrac{1}{16}w(90 - 2x)^2$$

This reduces to
$$x^2 - 137x + 3087 = 0$$
Solving the quadratic,
$$x = \tfrac{137}{2} \pm \tfrac{1}{2}\sqrt{137^2 - 4 \times 3087} = 68.5 \pm 40.1 = 28.4$$
Try two end spans of a 28 ft 6 in. span, and a center span of 33 ft 0 in. The factored load is
$$3.0 \times 1.7 = 5.1 \text{ kips/ft}$$
The required M_p for the end span is
$$M_p = 0.086 \times 5.1 \times 28.5^2 \times 12 = 4275 \text{ kip-in.}$$
As a check, calculate M_p for the center span:
$$M_p = \tfrac{1}{16} \times 5.1 \times 33^2 \times 12 = 4165 \text{ kip-in.}$$
(Slightly less, as an exact optimum was not chosen.) The required plastic modulus is
$$Z = \frac{4275}{36} = 118.8 \text{ in.}^3$$

A beam size W 24 $\times$ 55, with $Z = 134$ in.^{3}, is still the least-weight selection (AISCM). Although, in comparison with Example 8.6, the shifting of supports has appreciably reduced the required Z from 131.6 to 118.8 in.^{3}, no weight saving results. This, of course, is due to the discontinuous nature of available W shapes in relation to their strength properties. In comparison with Example 8.2, plastic design has effected a savings of 19.1% in weight over the allowable stress design. Moreover, the required computational work has been greatly reduced and simplified. No general conclusions should be drawn from these comparisons.

8.6 PLASTIC DESIGN OF FRAMES

In Examples 8.6 and 8.7 the determination of the required plastic modulus (Z) was relatively simple. There was only one distribution of moment possible at failure for each span, and the span requiring the greatest Z determined the beam selection. The solution was easily made by simple statics.

In a frame it is usually more convenient to make the analysis by the mechanism method, and there are more ways than one in which a possible failure mechanism can develop. Each possible mechanism will, in general, account for a different load at failure, only the lowest of which is correct and which thus provides the criterion for member selection. One recalls that "a chain is no stronger than its weakest link." One way to determine the correct mechanism is to try all possibilities, but as the number of members in a frame increases, this procedure gets increasingly cumbersome. As an alternative, after some experience with similar cases, if one can guess the correct mechanism in advance, its correctness can be verified by applying equations of static equilibrium to the elements between hinges as a preliminary to the construction of the bending moment diagram for the complete frame. If the correct mechanism has been chosen, the moment at each plastic hinge will be exactly equal to the plastic moment of the corresponding member. If the mechanism is incorrect, too great a load will have been calculated, and the bending moment at one or more plastic hinge locations will exceed the plastic moment capacity of the member. In any case, whether one tries all mechanisms or guesses the correct one, the foregoing moment check should be made to ensure the correctness of the solution. In very complex cases the aid of a computer can be sought, programmed to search out the lowest load corresponding to all possible mechanism solutions.

In applying the mechanism method, the method of virtual displacements is especially appropriate. If a system is in equilibrium, the total work done during an incremental displacement is zero; that is, the positive (external) work done by the applied forces is equal in magnitude to the negative (internal) work done. In the case of a plastic mechanism at failure load, the internal work is equal to the sum of the plastic moments multiplied by the rotations at the corresponding plastic hinges.

To illustrate the mechanism-virtual-displacement method, we redo the case shown in Fig. 8.5, the interior span of a uniformly loaded continuous beam. It is convenient to show the incremental displacement as if it origi- nated from the undeformed structure, as shown in Fig. 8.6.

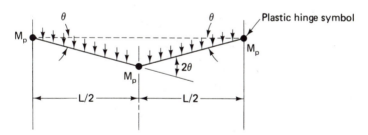

Fig. 8.6 Beam mechanism for interior span of a continuous beam.

The external work is equal to the total uniform load times the average deflection, which is $\theta L/4$:

$$\text{external work} = wL \times \frac{\theta L}{4}$$

The internal work is equal to the sum of the products of the plastic hinge moments multiplied by the incremental hinge rotation θ at the two ends and by 2θ at the center.

$$\text{internal work} = M_p(\theta + 2\theta + \theta) = 4M_p\theta$$

Equating the external and internal work we obtain, as before, the same Eq. (8.2):

$$M_p = \frac{wL^2}{16}$$

We now apply the same procedure to the simple hinged-base frame shown in Fig. 8.7, loaded horizontally at the top by force $P/3$ and vertically midspan by P. There are three possible mechanisms illustrated in Fig. 8.7: beam (a), panel (b), and a combination of the two (c).

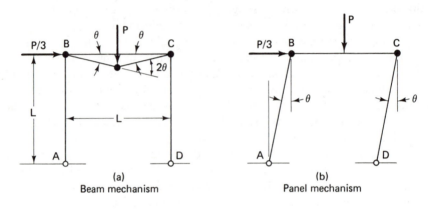

(a)
Beam mechanism

(b)
Panel mechanism

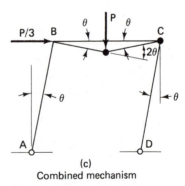

(c)
Combined mechanism

Fig. 8.7 Possible failure mechanisms for a single-span hinged-base bent.

For the beam mechanism, Fig. 8.7(a), equating external and internal work,

$$\frac{PL\theta}{2} = 4M_p\theta \qquad \text{or} \qquad M_p = \frac{PL}{8}$$

The mechanism shown in Fig. 8.7(b) is for failure by sidesway, usually termed a panel mechanism. In this case,

$$\frac{PL\theta}{3} = 2M_p\theta \quad \text{and} \quad M_p = \frac{PL}{6}$$

For the combined mechanism, Fig. 8.7(c), both the horizontal and vertical external forces do work and

$$\frac{PL\theta}{3} + \frac{PL\theta}{2} = M_p(2\theta + 2\theta)$$

whence

$$M_p = \frac{5}{24}PL$$

The largest value of M_p corresponds to the least load for a given M_p; hence the combined mechanism is the predicted mode of failure.

We now make a moment diagram check. The vertical force at D is obtained by taking moments about A for the entire frame. The horizontal shear at D is obtained by taking moments about C for member CD; that is, the base shear at D is equal to M_p/L, or $5P/24$. Thus the base shear at A must be $P/8$, and the moment at B is now determined as $PL/8$. The complete moment diagram can now be constructed as shown in Fig. 8.8, and the mo-

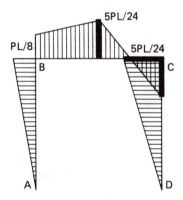

Fig. 8.8 Moment-diagram check on mechanism solution.

ment at the center of the beam is found by statics to be $5PL/24$, which is also the previously predicted value of M_p obtained by the combined mechanism shown in Fig. 8.7(c); thus the correctness of this solution is confirmed. Had this solution been incorrect, the moment at the center of the beam would have turned out to be greater than $5PL/24$.

Example 8.8

In Fig. 8.7, assume that $L = 14$ ft 0 in. and $P = 90$ kips. Select member sizes, assuming lateral support for buckling in the weak (out-of-plane) direction. Assume that the lateral load of $P/3$ is to be treated as equivalent to an earthquake shock load. Use A36 steel. (P also equals column load in design formulas.)

Solution

According to AISCS, Sec. 2.1, the load factor of 1.7 is to be applied for dead and live loads, and a load factor of 1.3 is to be applied when all loads, including earthquake, are considered. Checking the combined mechanism, with only a vertical load applied, we find that $M_p = PL/8$, identically same as for the beam mechanism of Fig. 8.7(a). No moment check is needed.

Calculate the required M_p:

Dead load plus live load:

$$M_p = \frac{1.7 \times 90 \times 14}{8} = 267.75 \text{ kip-ft}$$

For all loads, including earthquake:

$$M_p = \frac{1.3 \times 5 \times 90 \times 14}{24} = 341.25 \text{ kip-ft} \qquad \text{(controls)}$$

The required plastic modulus is

$$Z = \frac{341.25 \times 12}{36} = 113.75 \text{ in.}^3$$

Try a W 24 × 55:

$$Z = 134 \text{ in.}^3$$

$$A = 16.2 \text{ in.}^2$$

Check the width/thickness ratio:

$$\frac{b_f}{2t_f} = 6.9 < 8.5 \qquad \text{OK} \quad \text{(AISCS, AISCM)}$$

Check the depth/thickness ratio. Determine the vertical reaction at D by taking moments of forces on the complete structure about A.

$$V_D = \frac{30 \times 14 + 90 \times 7}{14} = 75 \text{ kips}$$

max. column load $= 1.3 \times 75 = 97.5$ kips

$$\frac{P}{P_y} = \frac{97.5}{36 \times 16.2} = 0.17 \qquad \text{[AISCS Formula (2.7-1a) applies]}$$

limiting $\dfrac{d}{t} = \dfrac{412}{6}(1 - 1.4 \times 0.17) = 52.3$

$$\frac{d}{t} = 59.7 > 52.3 \qquad \text{NG}$$

Try a section with a smaller depth to reduce d/t. Try a W 21 × 57:

$$Z = 129 \text{ in.}^3 \qquad A = 16.7 \text{ in.}^2 \qquad r_x = 8.36 \text{ in.} \qquad r_y = 1.35 \text{ in.}$$

$$\frac{b_f}{2t_f} = 5.0 < 8.5 \qquad \text{OK}$$

$$\frac{P}{P_y} = \frac{97.5}{36 \times 16.7} = 0.162$$

$$\text{limiting } \frac{d}{t} = \frac{412}{6}(1 - 1.4 \times 0.162) = 53.1$$

$$\frac{d}{t} = 52.0 < 53.1 \qquad \text{OK}$$

Check for the combined column load and bending moment [AISCS Formulas (2.4-2) and (2.4-3)].

To determine P_e, K (effective length factor) must be evaluated (refer to Section 4.3). Figure 4.4 will be used.

at the column top ($I_c/L_c = I_g/L_g$), $G = 1.0$

at the column bottom, for a hinged base, assuming that $G = 10.0$, then

$$K = 1.87 \qquad \text{(by Fig. 4.4)}$$

$$\frac{KL}{r_x} = \frac{1.87 \times 14 \times 12}{8.36} = 37.58$$

$$F'_e = 105.8 \text{ ksi (AISCS, Appendix A,Table 9)}$$

$$P_e = \frac{23}{12} AF'_e = \frac{23}{12} \times 16.7 \times 105.8 = 3386 \text{ kips (AISCS, Sec. 2.4)}$$

$$F_a = 19.38 \text{ ksi} \qquad \text{(AISCS, Appendix A, Table 3-36)}$$

$$P_{cr} = 1.7 \times 16.7 \times 19.38 = 550 \text{ kips} \qquad \text{[AISCS Formula (2.4-1)]}$$

Check by AISCS Formula (2.4-2):

$$\frac{97.5}{550.2} + \frac{0.85 \times 341.2 \times 12}{(1 - 97.5/3386) \times 129 \times 36} = 0.95 < 1 \qquad \text{OK}$$

Note: If the columns had proved to be inadequate, they could have been increased in size without changing the size of the beam.

Lateral Bracing Requirements

Plastic hinges may form at either B or C, and at the center. Assume that bracing is to be supplied at these three locations and, in addition at beam $\frac{1}{4}$ points. Considering the moment diagram in Fig. 8.8 and the required braced distance from C to the left, check the adequacy of $\frac{1}{4}$-point bracing.

$$\frac{M}{M_p} = 0 \quad \text{and AISCS Formula (2.9-1a) applies}$$

$$\frac{l_b}{r_y} = \frac{3.5 \times 12}{1.35} = 31$$

$$\frac{l_{cr}}{r_y} = \frac{1375}{36} + 25 = 63.2 > 31 \quad \text{OK}$$

For bracing in column, adjacent to hinge at C (or B for reversed direction of horizontal force), M/M_p will be between -0.5 and -1.0 and AISCS Formula (2.9-1b) will apply:

$$\frac{l_{cr}}{r_y} = \frac{1375}{36} = 38.2, \text{ and } l_{cr} = 38.2 \times 1.35 = 51.6 \text{ in.}$$

To minimize the required bracing between the upper column brace point and the column base, use the full allowable 51 in. (4.25 ft). To check bracing requirements in the remainder of the column, refer to AISCS, Sec. 2.9: "in regions not adjacent to a plastic hinge, the maximum distance between points of lateral support shall be such as to satisfy the requirements of Formula (1.5-6a), (1.5-6b), or (1.5-7) as well as Formulas (1.6-1a) and (1.6-1b) in Part I of this specification. For this case the values of f_a and f_b shall be computed from the moment and axial force at factored loading, divided by the applicable load factor."

The moment at the upper brace point, 4.25 ft below C in Fig. 8.8 is

$$M = \frac{9.75}{14} \times 341.25 = 237.7 \text{ kip-ft}$$

column load $= 97.5$ kips

Divide by the load factor of 1.3:

$$M = \frac{237.7}{1.3} = 182.8 \text{ kip-ft}$$

$$P = \frac{97.5}{1.3} = 75 \text{ kips}$$

Check for no bracing at all in the lower 9 ft 9 in.

$$C_b = 1.75 \quad \text{(AISCS, Sec. 1.5.1.4.6a, with } M_1 = 0)$$

Try Formula (1.5-7) ($d/A_f = 4.94$, AISCM):

$$F_b = \frac{12 \times 1000 \times 1.75}{9.75 \times 12 \times 4.94} = 36.3 \text{ ksi} > 22$$

$$= 22 \text{ ksi}$$

The section modulus $S_x = 111$ in.^{3} (AISCM):

$$f_b = \frac{182.8 \times 12}{111} = 19.76 \text{ ksi} < 22 \quad \text{OK}$$

Check the 9 ft-9 in. unbraced column segment for the combined stress criteria of AISCS, Sec. 1.6.1. Assume that $K = 1$.

$$C_m = 0.85$$

$$\frac{L}{r_y} = \frac{9.75 \times 12}{1.35} = 86.7 \quad \text{(governs)}$$

$$\frac{L}{r_x} = \frac{14.0 \times 12}{8.36} = 20.1$$

$$F_a = 14.6 \text{ ksi} \quad \text{(AISCS, Appendix A, Table 3-36)}$$

$$f_a = \frac{75}{16.7} = 4.49 \text{ ksi}$$

$$f_{bx} = 19.76 \text{ ksi} \quad \text{(previously calculated)}$$

$$\frac{f_a}{F_a} = 0.308 > 0.15 \quad \text{[AISCS Formulas (1.6-1) apply]}$$

Check by AISCS Formula (1.6-1a). $F'_{ex} = 762$ ksi.

$$\frac{4.49}{14.6} + \frac{0.85 \times 19.76}{(1 - 4.49/762)22} = 1.08 > 1 \quad \text{NG}$$

In lieu of changing the column size, add a second brace that meets the plastic design criteria of AISCS Formula (2.9-1b), an additional 51 in. down the column, reducing f_b at the brace. Calculate the moment at the brace and check for $l = 5$ ft 6 in.

$$M = \frac{5.5}{14} \times 341.25 = 134 \text{ kip-ft and reduced by } LF$$

$$= \frac{134.1}{1.3} = 103.2 \text{ kip-ft}$$

$$f_{bx} = \frac{103.2 \times 12}{111} = 11.56 \text{ ksi}$$

$$\frac{l}{r_y} = \frac{5.5 \times 12}{1.35} = 48.9$$

$$F_a = 18.45 \text{ ksi} \quad \text{(AISCS, Appendix A, Table 3-36)}$$

Check by AISCS Formula 1.6-1a:

$$\frac{4.49}{18.45} + \frac{0.85 \times 11.56}{\simeq 21.9} = 0.69 < 1 \quad \text{OK}$$

Note that the lateral braces must hold the members against both lateral movement and twist. Design of these details is omitted.

At the upper corner knee connections, where a plastic hinge may form, the full plastic moment creates high shear in the web at the corner. The full yield strength of the flange areas of the beam must be transmitted by shear into the column web, and vice versa (see AISCS Commentary, Sec. 2.5). If the web thickness is inadequate, a diagonal stiffener may be added, as shown. The design will be made for the full-capacity hinge moment of $36 \times 129 = 4644$ kip-in.

Calculate the shear in the web:

$$V = \frac{4644}{0.95 \times 21.06} = 232 \text{ kips} \quad \text{(AISCS Commentary, Sec. 2.5)}$$

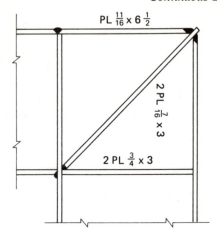

The shear capacity of the web at yield is

$$0.55 \times 36 \times 21.06 \times 0.405 = 169 \text{ kips} \qquad \text{(inadequate)}$$

Add a diagonal stiffener to take up the deficiency in web shear capacity. Calculate the force in the stiffener:

$$F = \frac{232 - 169}{0.707} = 89.1 \text{ kips}$$

The stiffener area required is

$$A = \frac{89.1}{36} = 2.48 \text{ in.}^2$$

Try two PL $\frac{7}{16} \times 3$:

$$A = 2.63 \text{ in.}^2$$

Check the b/t ratio:

$$\frac{b}{t} = \frac{3}{0.4375} = 6.86 < 8.5 \qquad \text{OK}$$

Add end plate and vertical stiffeners to transmit column flange forces into beam web. The column flange area is $6.56 \times 0.65 = 4.26$. The end plate to be $\frac{11}{16} \times 6\frac{1}{2}$:

$$A = 4.47 \text{ in.}^2$$

The vertical stiffeners are to be two PL $\frac{3}{4} \times 3$:

$$A = 4.5 \text{ in.}^2$$

Welds attaching the end plates and vertical stiffeners to the beam web should be sized so as to fully develop the web shear capacity of 169 kips. Weld stresses at factored loads are to be 1.7 times the allowable-stress design values of AISCS, Table 1.5.3, or this text, Table 6.1. Weld design involves elementary procedures covered in Chapter 6 and will not be carried further in this example.

The plastic design of a frame having the same dimensions and load requirements of Example 8.5 will now afford a further opportunity to compare weights between a plastic and an allowable stress design. In the absence of lateral load at the top, it may be presumed that failure will be by beam mechanism.

Example 8.9

Same as Example 8.5, but by plastic design.

Solution

Beam selection will be on the basis illustrated in Fig. 8.5, with any one of the three beam spans requiring the same plastic moment,

$$M_p = \frac{wL^2}{16} = \frac{1.7 \times 3 \times 25^2}{16} = 199.2 \text{ kip-ft}$$

The required plastic modulus is

$$Z = \frac{199.2 \times 12}{36} = 66.4 \text{ in.}^3$$

Try a W 18 × 35, with

$$Z = 66.5 \text{ in.}^3 \qquad A = 10.3 \text{ in.}^2$$

Check the adequacy of the same section as column element *AB* or *EF*. As in allowable stress design, the framed columns must provide added stability for the interior columns, which are hinged, top and bottom, and not continuous with the main frame. End columns will be designed for half of the full load on all three spans,

$$P = \frac{1.7 \times 3 \times 75}{2} = 191.25 \text{ kips}$$

Check the width/thickness ratios by AISCS, Sec. 2.7:

$$\frac{b}{2t_f} = 7.1 < 8.5 \qquad \text{OK}$$

$$\frac{d}{t} = 59.0$$

Check the adequacy by AISCS Formula (2.7-1a) or (2.7-1b).

The actual factored load, $P = 1.7 \times 3 \times 12.5 = 63.75$ kips, is used as in Example 8.5.

$$\frac{P}{P_y} = \frac{63.75}{36 \times 10.3} = 0.172 < 0.27 \qquad \text{AISCS Formula (2.7-1a) applies.}$$

The limiting d/t_w, by AISCS Formula (2.7-1a), is

$$\frac{412}{\sqrt{36}}(1 - 1.4 \times 0.172) = 52.1 < 59.0 \qquad \text{NG}$$

The following sections were tried: W 16 × 45, W 18 × 50, and W 21 × 50. These all met the d/t_w limitation but failed in the combined axial load and bending moment check by AISCS Formula (2.4-2). These calculations are omitted since they are essentially similar to the following.

Try a W 21 × 57:

$$A = 16.7 \text{ in.}^2 \qquad Z = 129 \text{ in.}^3 \qquad I = 1170 \text{ in.}^4$$

$$r_x = 8.36 \text{ in.} \qquad \frac{d}{t_w} = 52.0 \quad \text{(AISCM)}$$

$$\frac{b_f}{2t_f} = 5.0 < 8.5 \qquad \text{OK} \quad \text{(AISCS, Sec. 2.7)}$$

Check d/t_w:

$$\frac{P}{P_y} = \frac{63.75}{36 \times 16.7} = 0.106$$

The limiting d/t_w, by AISCS Formula (2.7-1a), is

$$\frac{412}{\sqrt{36}}(1 - 1.4 \times 0.106) = 58.5 > 52.0 \qquad \text{OK}$$

Determine the effective length factor, K (Section 4.3).

$$\frac{I_c}{L_c} = \frac{1170}{18 \times 12} = 5.42$$

$$\frac{I_g}{L_g} = \frac{510}{25 \times 12} = 1.70$$

$$G_A = \frac{5.42}{1.70} = 3.19$$

$$G_B = 10 \qquad \text{(hinge)}$$

$$K = 2.3 \qquad \text{(from Fig. 4.4)}$$

$$\frac{KL}{r_x} = \frac{2.3 \times 18 \times 12}{8.36} = 59.4$$

$$F_a = 17.49 \text{ ksi} \qquad \text{(AISCS, Appendix A, Table 3-36)}$$

$$F'_{ex} = 42.33 \text{ ksi} \qquad \text{(AISCS, Appendix A, Table 9)}$$

$$P_{cr} = 1.7 \times 16.7 \times 17.49 = 496.5 \text{ kips}$$

$$P_{ex} = \frac{23 \times 16.7 \times 42.33}{12} = 1355 \text{ kips}$$

$$M_m \equiv M_p = 129 \times 36 = 4644 \text{ kip-in.}$$

$$M = 66.4 \times 36 = 2390 \text{ kip-in.}$$

Check by AISCS Formula (2.4-2):

$$\frac{191.2}{496.5} + \frac{0.85 \times 2390}{(1 - 191.2/1355)4644} = 0.894 < 1.0 \qquad \text{OK}$$

Interior column design at locations C and D: these columns are independent of the main frame and the design is essentially the same as in Example 8.5. The plastic equalization of all beam end moments would result in a slightly smaller column load at working load, but the required column size is the same.

Although heavier end columns are required in the plastic design, the overall weight of the main-frame material is 4677 lb, 873 lb less than the 5550 lb

required for the elastic allowable stress design of Example 8.5. The 15.7%
overall saving is appreciable, especially if a number of frames are required.

Plastic design has been used in multistory buildings, most commonly
for the beams in either fully braced structures or those with relatively few
floors for which lateral forces of wind or earthquake are not a primary
consideration. Plastic design probably offers the greatest advantage in application to one-story frames similar to the case in Example 8.9. Its use will
increase as an adjunct and part of load and resistance factor design, which
will be previewed in Chapter 9. The greater simplicity of plastic design analysis
procedures in comparison with continuous beam analysis was well demonstrated by Examples 8.5 and 8.9. For those who wish to pursue the subject
further, ASCE *Manual 41*, "Plastic Design in Steel—A Guide and Commentary" is recommended. An early text on the subject, *Plastic Design of
Steel Frames*, by Lynn S. Beedle, (John Wiley & Sons, Inc., New York, 1958)
also provides many detailed analysis and design solutions of a variety of
frames.

PROBLEMS

8.1. Similar to Example 8.1, but change to three spans of 36 ft, each designed for
a dead load of 1 kip/ft and a live load of 3 kip/ft. Use A36 steel. Assume
continuous lateral support. Use AISCM tables to calculate moments. Select
member sizes.

8.2. †Similar to Example 8.2, but change the center span to 40 ft, the end spans
to 34 ft each. Analysis is to be by moment distribution. The live load is 3
kip/ft, the dead load 1 kip/ft. Use A36 steel.

8.3. Similar to Example 8.3, but end spans of 34 ft with cantilever overhang of
5 ft, leaving a simply supported suspended center beam of 30 ft. Use A36
steel. Same loads as Problems 8.1 and 8.2. Compare total weights for the three
alternative solutions.

8.4. †Select column and beam sizes for allowable stress design for the two-span
frame shown. Use A36 steel. The dead load is 1 kip/ft, the live load, 2 kip/ft.
Compare the total weight with that of Example 8.5, which has the same
loads and overall length. Lateral support is assumed for the end columns
and the beam.

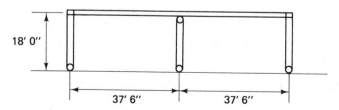

†If used as assigned, Problems 8.2, 8.4, and 8.10 could be given a grade weighting factor
of 3×.

8.5. Determine the plastic shape factor for the box section shown. Check the suitability of the section for plastic design using steels with yield points of 36 and 50 ksi.

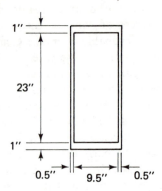

8.6. Show that $M_p = 0.086wL^2$ and that the plastic hinge for positive moment occurs at a distance $0.414L$ from A at the left end of the simple support. Same conditions as in Fig. 8.4. (*Hint:* From the equilibrium of the entire span, write an expression for R_A. Then write a general expression for M at any distance x from A. Set $dM/dx = 0$ for maximum M.)

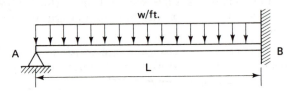

8.7. Using the mechanism method, determine an expression for M_p. Compare with results of Problem 8.6.

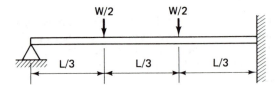

8.8. Same conditions as Problem 8.1, but by plastic design.

8.9. Same conditions as Problem 8.2, but by plastic design. Compare the total weights of the results of Problems 8.1, 8.2, 8.3, 8.8, and 8.9.

8.10. ‡Design the frame as shown by plastic design. Use A36 steel. The lateral load

‡Grade weighting factor of $3 \times$.

of 10 kips is to be treated as equivalent to an earthquake load. Determine the required locations for lateral support. Design a corner knee connection.

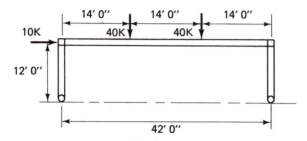

8.11. Same as Problem 8.4, but by plastic design. Compare the total weight with the results of Problem 8.4.

8.12. In Example 8.9, show that if one end span carries full live load, the frame is no more likely to fail by sidesway than by the beam mechanism that was presumed to be critical.

9

LOAD AND RESISTANCE FACTOR DESIGN

9.1 INTRODUCTION

Looking to the future, load and resistance factor design when fully implemented will provide a rational and optimal approach toward achieving the paramount qualities of safety and economy in design. The reader should review Section 1.4, which covers the development of *allowable stress design* (ASD), *plastic design* (PD), and *load and resistance factor design* (LRFD).†
In addition, Sections 1.5 and 1.6 discuss safety and economy, both of which involve decisions about design, fabrication, and erection. The requirements of safety form the limits that must not be violated, and the requirements of economy provide the optimal solution within these limits.

Structural safety in design is accomplished by assuring, by means of design calculations, that the limits of structural usefulness given by the applicable structural specification, such as the AISCS, are not violated. In its Preface, the AISCS cautions that "independent professional judgment must be exercised" in applying the specification and that "it is not intended to cover the infrequently encountered problems within the full range of structural design practice. . . ." The AISCS is regularly up-dated to include new developments in research and practice and its writers are especially concerned with safety. In 1978 the AISCS was composed of two parts: Part 1 provided rules for allowable stress design (ASD), and Part 2 defined the criteria for plastic design (PD). Part 3, to be added in 1979 or 1980, will

†Also termed "Limit States Design" by many.

270

define the criteria for LRFD, a procedure that aims to make full use of available test information, design experience, and engineering judgment, applied by use of probabilistic analyses.

In ASD the limits of structural usefulness are allowable stresses that must not be exceeded when the forces in the steel structure are determined by an elastic analysis. The allowable stresses F_{all} are defined by the relationship

$$F_{all} = \frac{F_{lim}}{FS} \qquad (9.1)$$

where FS is the factor of safety, and F_{lim} is a stress that denotes a limit of usefulness such as the yield stress F_y, a critical (buckling) stress F_{cr} (column stability, beam stability, or plate stability), the tensile stress F_u at which the member fractures, or the stress range F_{sr} in fatigue. For example, the allowable stresses in tension (AISCS, Sec. 1.5.1.1) are the smaller of $0.6F_y$ (limit state of yielding, $FS = 1/0.6 = 1.667$) and $0.5F_u$ (limit state of fracture, $FS = 1/0.5 = 2.0$). The actual stresses that must not exceed the allowable stresses are determined by elastic analysis for the working loads on the structure.

In Part 2 of the AISCS (PD), the limit of structural usefulness is a load P_u which will cause a plastic mechanism to form. This limiting load is then compared with the factored working loads:

$$(LF)P_w \leq P_u \qquad (9.2)$$

where P_w represents the working loads and LF is a load factor. (LF = 1.7 for gravity loads and LF = 1.3 for combined gravity and wind or earthquake loads).

The general design criteria are illustrated in Fig. 9.1, where R represents the resistance (strength) of a structural element and Q is the load effect (calculated force due to the maximum loads expected during the life of the structure). The symbols Q_s and R_s represent, respectively, the load effect due to the specified working loads and the minimum specified resistance. In Fig. 9.1(a) the picture represents ASD, using FS, and in Fig. 9.1(b) we see the role of LF in PD. Both the factor of safety and the load factor serve the purpose of providing a margin of safety between R_s and Q_s to take care of the unforeseen but possible eventuality that the actual load might exceed the specified value and/or that the actual resistance is smaller than the specified value. These uncertainties are in the nature of loads and resistances. In fact, we can easily visualize that both load effects and resistances have a form of a probabilistic distribution [Fig. 9.1(c)], characterized by a bell-shaped curve that has a mean value (R_m or Q_m) and a standard deviation. Exceedance of a limit state is, then, the condition that $R < Q$, and this is always possible. Structural safety is thus defined as the acceptably small

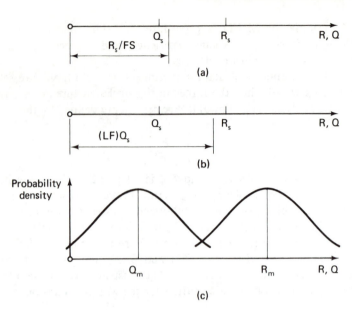

Fig. 9.1 Definitions of structural safety.

probability of Q exceeding R, and the true role of FS and LF is to assure that this probability is negligibly small.

9.2   PROBABILISTIC DEFINITION OF STRUCTURAL SAFETY

One could start the determination of the required margin of safety by stipulating an acceptably small probability of exceeding a limit of structural usefulness (*limit state*), and then from the known probabilistic distributions of R and Q one could, by the calculus of probability theory, arrive at the appropriate margin of safety. This is how the various factors were arrived at for the newly proposed design method (see Ref. 9.1 for the current version of the LRFD criteria).

The method of arriving at a probabilistic safety margin is as follows. A structure is safe (i.e., a limit state is not violated) if $R - Q \geq 0$, or $R/Q \geq 1$, or $ln\, R/Q \geq 0$. The distribution of $ln\, R/Q$ is shown in Fig. 9.2. The limit state is violated if $ln\, R/Q$ is negative, and the probability of this happening is represented as the shaded area in Fig. 9.2. The smaller this area is, the more reliable is the structural element. The shaded area varies in size as the distance of the mean value of $ln\, R/Q$ from the origin (Fig. 9.2). This

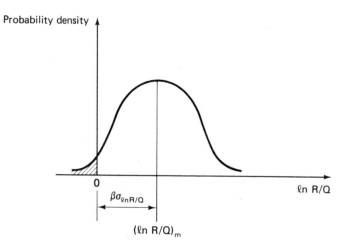

Fig. 9.2 Definition of the safety index.

distance depends on two factors: the width of the distribution curve, as characterized by its standard deviation $\sigma \, ln \, R/Q$ (i.e., the scatter of the data which make up the distribution), and a factor β, which is called the *safety index*. For any given distribution $ln \, R/Q$, the larger β is, the smaller is the probability of exceeding a limit state. In fact, if we know the exact distribution curve, we can directly relate this probability to the safety index β.

Unfortunately, the probabilistic distributions of R and Q are known only for very few resistance and load components, but at least we know the mean values and the standard deviation from the analysis of data on loads and material properties. From this knowledge, and from some approximations described in Ref. 9.2, we can obtain the following simple formula for the safety index β:

$$\beta = \frac{ln \, (R_m/Q_m)}{\sqrt{V_R^2 + V_Q^2}} \tag{9.3}$$

In this equation the terms R_m and Q_m are the mean values of the resistance R and the load effect Q, respectively, and V_R and V_Q are the corresponding coefficients of variation ($=$ standard deviation/mean).

We can now find out the value of β inherent in, say, Part 1 of the AISCS by designing an element, then obtaining the appropriate statistical data, and so computing the value of β. This process is called *calibration* (Ref. 9.3).

Let us take, for example, a simply supported beam of 30-ft length, supporting a specified dead load of 50 lb/ft² and a specified office live load of 50 lb/ft². The beams are spaced 20 ft apart and they are fully braced along their entire length.

The tributary area is

$$A_T = 30 \times 20 = 600 \text{ ft}^2$$

The live-load reduction factor (Ref. 9.2) is the smaller of

$$0.0008 A_T = 0.0008 \times 600 = 0.48$$

and

$$0.23\left(1 + \frac{D}{L}\right) = 0.23\left(1 + \frac{50}{50}\right) = 0.46$$

$$w = 20 \times [50 + 50(1 - 0.46)] \times \frac{1}{1000} = 1.540 \text{ kips/ft}$$

The maximum moment

$$M = \frac{wL^2}{8} = \frac{1.54 \times 30^2 \times 12}{8} = 2079 \text{ kip-in.}$$

Using Part 1, AISCS, and compact sections, $F_b = 0.66 F_y$. The required elastic section modulus for $F_y = 36$ ksi is

$$S = \frac{M}{F_b} = \frac{2079}{0.66 \times 36} = 87.5 \text{ in.}^3$$

A W 18 $\times$ 50 beam is required ($S = 88.9 \text{ in.}^3$).

For calibration we need to know the statistics of the load effect Q and the resistance R. The load effect is equal to the maximum moment. The distributed load is composed of the dead and the live loads. Research leading up to the development of LRFD criteria (Refs. 9.1 and 9.3) has shown that the mean dead load is equal to the nominal load:

$$D_m = D = 50 \text{ lb/ft}^2 \qquad \text{and} \qquad V_D = 0.06$$

and the lifetime maximum mean office live load is

$$L_m = 14.9 + \frac{763}{\sqrt{2A_T}} = 14.9 + \frac{763}{\sqrt{2 \times 600}} = 37 \text{ lb/ft}^2 \qquad V_L = 0.24$$

The mean distributed load is thus equal to

$$w_m = (w_D)_m + (w_L)_m = 20(50 + 37)\frac{1}{1000} = 1.740 \text{ kips/ft}$$

so

$$M_m = Q_m = \frac{wL^2}{8} = \frac{1.740 \times 30^2 \times 12}{8} = 2349 \text{ kip-in.}$$

The coefficient of variation of Q is

$$V_Q = \sqrt{\frac{(D_m V_D)^2 + (L_m V_L)^2}{(D_m + L_m)^2}} = \sqrt{\frac{(50 \times 0.06)^2 + (37 \times 0.24)^2}{(50 + 37)^2}} = 0.11$$

The resistance is equal to the plastic moment:

$$R_m = (M_p)_m = (F_y)_m (Z)_m$$

The data on material properties and section properties (Ref. 9.3) give

$$(F_y)_m = 1.05\,(F_y)_n \qquad V_{F_y} = 0.10$$

$$(Z)_m = (Z)_n \qquad V_Z = 0.05$$

where the subscript m defines mean values and the subscript n nominal values, respectively.

$$(F_y)_n = 36\text{ ksi} \qquad (Z)_n = 101\text{ in.}^3$$

$$R_m = 1.05 \times 36 \times 101 = 3818\text{ kip-in.}$$

$$V_R = \sqrt{V_{F_y}^2 + V_Z^2} = \sqrt{0.1^2 + 0.05^2} = 0.11$$

Thus

$$\beta = \frac{ln\,(R_m/Q_m)}{\sqrt{V_R^2 + V_Q^2}} = \frac{ln\,(3819/2349)}{\sqrt{0.11^2 + 0.11^2}} = 3.12$$

In a similar manner, calibrations were performed for many types of structural elements (Refs. 9.3–9.5) and on the basis of these calibrations, $\beta = 3.00$ for members and $\beta = 4.50$ for connections were selected for the LRFD criteria. Since the factors of safety in the AISC Specification resulted from an evolutionary process of many years of experience, there was some scatter in the β's. The new LRFD method removes this scatter, thereby ensuring uniform reliability at least equal to the accepted designs in the current AISCS. One advantage of probability-based design is therefore that a more uniform reliability results. The other advantage is that the variability of the various design elements can be properly accounted for, as will be seen in the next section.

9.3 LOAD AND RESISTANCE FACTOR DESIGN

In the proposed LRFD criteria (Ref. 9.1) the designer is not expected to manipulate statistical data and compute β values for comparison with prescribed values, as illustrated by the example in the previous section. Instead, he proceeds as in the current AISCS, following prescribed rules in determining resistances and using multiple load factors. The design check to be made is as follows:

$$\gamma_A \sum_{i=1}^{w} \gamma_i Q_i \le \phi R_n \qquad (9.4)$$

In this formula γ_A is an analysis factor, accounting for the uncertainties of the structural analysis, γ_i's are load factors by which the individual load effects Q_i are multiplied to account for the uncertainties of the loads, R_n is a nominal resistance (say $R_n = M_p = F_y Z$), and ϕ is a resistance factor that accounts for the uncertainties inherent in the determination of the resistance. The γ's are larger than unity and ϕ is less than unity.

The γ and ϕ factors are determined from the specified safety index β and the appropriate statistical data, using the expressions (Ref. 9.3)

$$\phi = \frac{R_m}{R_n} e^{-0.55\beta V_R} \tag{9.5}$$

$$\gamma_A = e^{0.55\beta V_A} \tag{9.6}$$

$$\gamma_D = 1 + 0.55\beta V_D \tag{9.7}$$

$$\gamma_L = 1 + 0.55\beta V_L \tag{9.8}$$

For example, for beams under dead and live loads,

$$\phi = 0.86 \qquad \gamma_A = 1.1 \qquad \gamma_D = 1.1 \qquad \gamma_L = 1.4$$

and the design criterion then becomes

$$1.1(1.1\,M_D + 1.4\,M_L) \le 0.86 M_u \tag{9.9}$$

where M_D and M_L are the moments due to dead and live loads, respectively, and M_u is the nominal ultimate moment of the beam. In Eq. (9.9) we can see the advantage of LRFD over ASD or PD: the smaller load factor for dead loads reflects the fact that the determination of dead loads is more certain than that of live loads. Thus where dead load dominates, this fact is taken into account in design.

At the time of writing of this chapter, LRFD was not yet part of AISCS, but an unofficial draft of the criteria had been published (Ref. 9.1) together with the background (Refs. 9.3–9.8). The reader should seek to learn from AISC the current status of adoption. The initial presentation of LRFD criteria will be in the *AISC Engineering Journal*, where preliminary papers have already appeared (Refs. 9.6–9.9). After discussion and trial in design offices it will be adopted as an alternative method in AISCS. Some coefficients will be changed from those in Ref. 9.1, but the essence of the method, that is, the use of resistance factors and multiple load factors, has been well established and is already in use in many countries.

9.4   ILLUSTRATIVE EXAMPLES

Following are five examples illustrating the LRFD method of designing steel structures. Since LRFD may not yet have become a part of AISCS, and since the reader may not have access to Ref. 9.1, the LRFD requirements needed for each of the following examples are given after the problem statement. If new AISCS LRFD criteria are available, the revision of the following examples could be suitable problem assignments.

Example 9.1 is the design of a series of simply supported beams. Example 9.2 is a continuous beam designed for two loading conditions: under final service conditions and under construction loading. This problem makes use of plastic design with investigation of the limit states of lateral-torsional buckling (LTB), flange local buckling (FLB), and web local buckling (WLB). The bases for the formulas for the ultimate moment M_u and the slenderness limits are the same as in the AISCS; only the discontinuities in the AISCS are removed, and the critical lateral-torsional buckling moment is based on a more exact formula.

The elastic lateral-torsional buckling moment of a simply supported wide-flange beam bent by end moments about the major axis is (Ref. 1.6)

$$M_{cr} = \frac{C_b \pi}{L_b} \sqrt{EI_y GJ} \sqrt{1 + \frac{\pi^2 E C_w}{GJL_b^2}} \tag{9.10}$$

where

C_b is defined in Sec. 3.4 (and in Flowchart 9.1)
L_b = unbraced length
E = elastic modulus (E = 29,000 ksi for steel)
G = shear modulus (G = 0.385E for steel)
I_y = weak axis moment of inertia
J = torsion constant
C_w = warping constant

If we substitute the relationships

$$C_w = \frac{I_y(d - t_f)^2}{4} \qquad I_y = Ar_y^2 \qquad \lambda_b = \frac{L_b}{r_y}$$

we obtain the expression for M_u at the bottom of Flowchart 9.1. Elastic buckling ceases to govern when the critical stress (Fig. 9.3)

$$f_{cr} = \frac{M_{cr}}{S_x} = F_y - F_r$$

where F_r is the residual stress. By assuming an average residual stress of F_r = 10 ksi, the moment that limits elastic buckling is $M_r = S_x(F_y - 10 \text{ ksi})$, as shown in Fig. 9.3. The corresponding unbraced length L_{br} (i.e., $\lambda_{br} = L_{br}/r_y$) is obtained by setting $M_{cr} = M_r$ in Eq. (9.9) and solving for L_{br}. This results in the equation for λ_{br} given in Flowchart 9.1.

Up to a slenderness value of λ_{bp}, the ultimate moment M_u is equal to the plastic moment M_p (Fig. 9.3), and between λ_{br} and λ_{bp}, M_u is assumed to vary as a straight line. The limiting slenderness values λ_{bp} are essentially the same as the compactness requirements of AISCS.

Example 9.3 illustrates the LRFD of a column, Example 9.4 designs a column with end moments, and Example 9.5 takes us through the design of a plate girder.

Example 9.1

A parallel row of simply supported steel wide-flange beams of 35-ft length spaced 15 ft apart must support an 8-in.-deep concrete slab and a live load of 100 lb/ft^2 in addition to its own weight. Design the beams using LRFD. The concrete slab provides continuous lateral support to the beams. Use A36 steel.

LRFD design requirements (Sec. 1.3.2, Ref. 9.1):

$$\phi M_p \geq \gamma_A(\gamma_D M_D + \gamma_L M_L) \quad \text{design condition}$$

$$M_p = ZF_y \quad \text{plastic moment}$$

$$Z = \text{plastic section modulus} \quad \text{(See Section 3.3)}$$

$$\gamma_A = 1.1 \quad \gamma_D = 1.1 \quad \gamma_L = 1.4 \quad \text{load factors}$$

$$M_D = \text{dead-load moment}$$

$$M_L = \text{live-load moment}$$

Beam design requirements (See. 2.3.3, Ref. 9.1):

$$\phi = 0.86 \quad \text{resistance factor}$$

compactness requirements: $\quad \dfrac{b_f}{2t_f} \leq \dfrac{65}{\sqrt{F_y}} \quad \text{and} \quad \dfrac{d}{t} \leq \dfrac{640}{\sqrt{F_y}}$

$$F_y = \text{yield stress, ksi}$$

Solution

Distributed dead load/beam:

Concrete (150 lb/ft^3): $\qquad$ $(\frac{8}{12} \times 1 \times 15)0.15 = 1.500$ kips/ft

Estimated weight of steel: $\qquad \qquad \underline{\qquad 0.100 \text{ kips/ft}}$

$$w_D = 1.6 \text{ kips/ft}$$

Maximum moment at center of beam ($M_{\max} = wL^2/8$):

$$M_D = \frac{1.6 \times 35^2 \times 12}{8} = 2940 \text{ kip-in.}$$

Distributed live load on beam ($w_L = 0.100 \times 15 = 1.5$ kips/ft):

$$M_L = \frac{1.5 \times 35^2 \times 12}{8} = 2756 \text{ kip-in.}$$

$$\gamma_A(\gamma_D M_D + \gamma_L M_L) = 1.1(1.1 \times 2940 + 1.4 \times 2756) = 7802 \text{ kip-in.}$$

$$\phi Z_x F_y \geq 7802 \text{ kip-in.}$$

$$Z_x \geq \frac{7802}{0.86 \times 36} = 252 \text{ in.}^3$$

From the Plastic Design Selection Table (AISCM), select W 24 × 94 as a trial section.

$$Z_x = 254 \text{ in.}^3 > 252 \text{ in.}^3 \qquad \text{OK}$$

$$\frac{d}{t} = 47.2 < \frac{640}{\sqrt{F_y}} = 106.7 \qquad \text{OK}$$

$$\frac{b_f}{2t_f} = 5.18 < \frac{65}{\sqrt{F_y}} = 10.83 \qquad \text{OK}$$

Use a W 24 × 94.

Example 9.2

Design a two-span continuous beam in an office building.

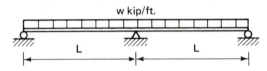

Span L = 30 ft.

Beam spacing : 20 ft.

Joist spacing : 2 ft. (spacing of braced joints along beam)

Loading: Mean dead load, exclusive of beam weight, w_D = 1.20 kip/ft.

The mean office live-load intensity (Sec. C1.3, Ref. 9.1) is

$$L = 14.9 + \frac{763}{\sqrt{A_I}} \qquad \text{lb/ft}^2$$

This is a reduced live load, and $A_I = 2A_T$, where A_I is the influence area and A_T is the tributary area.

$$A_T = 30 \text{ ft} \times 20 \text{ ft} = 600 \text{ ft}^2/\text{span}$$

$$L = 14.9 + \frac{763}{\sqrt{2 \times 600}} = 37 \text{ lb/ft}^2$$

$$w_L = 20 \times \frac{37}{1000} = 0.740 \text{ kip/ft}$$

In addition, consider a construction load, acting before joists provide lateral support, of a 5-kip concentrated load at the center of each span, plus the weight of the beam. Lateral support during construction is present only at the supports. Use A36 wide-flange beams.

LRFD design requirements (Ref. 9.1):
Load factors:

$$\gamma_A = 1.1 \qquad \gamma_D = 1.1 \qquad \gamma_L = 1.4 \qquad \gamma_C = 1.4$$

The moment capacity (Sec. 2.3.3, Ref. 9.1) is

$$\phi M_u \geq \gamma_A \sum \gamma_i M_i$$
$$\phi = 0.86$$

M_u is the lowest M_u from the following three limit-state moments:

1. Lateral-torsional buckling (LTB).
2. Flange local buckling (FLB).
3. Web local buckling (WLB).

Determination of M_u is made according to Flowcharts 9.1, 9.2, and 9.3 according to the general scheme shown in Fig. 9.3. This figure gives the variation of M_u with the slenderness parameters L_b/r_y (L_b is unbraced length, r_y is weak axis radius of gyration), $b_f/2t_f$, and d/t_w, representing the three limit states LTB, FLB, and WLB.

Fig. 9.3 Schematic variation of M_u with slenderness parameters λ_b.

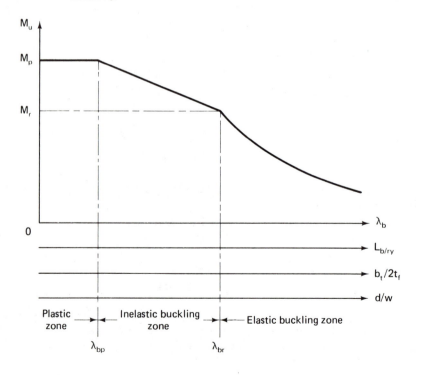

Flowchart 9.1

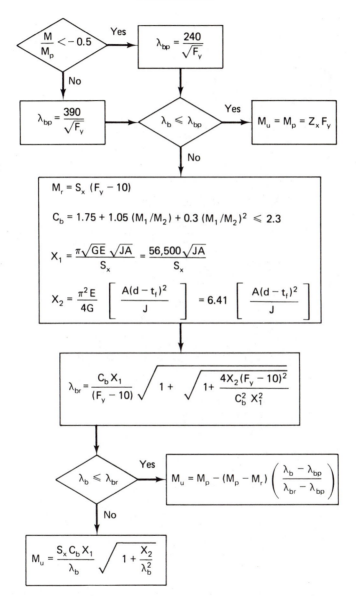

$\dfrac{M}{M_p} < -0.5$ — Yes → $\lambda_{bp} = \dfrac{240}{\sqrt{F_y}}$

No ↓

$\lambda_{bp} = \dfrac{390}{\sqrt{F_y}}$ → $\lambda_b \leq \lambda_{bp}$ — Yes → $M_u = M_p = Z_x F_y$

No ↓

$$M_r = S_x (F_y - 10)$$

$$C_b = 1.75 + 1.05 (M_1/M_2) + 0.3 (M_1/M_2)^2 \leq 2.3$$

$$X_1 = \frac{\pi\sqrt{GE}\ \sqrt{JA}}{S_x} = \frac{56{,}500\sqrt{JA}}{S_x}$$

$$X_2 = \frac{\pi^2 E}{4G} \left[\frac{A(d - t_f)^2}{J} \right] = 6.41 \left[\frac{A(d - t_f)^2}{J} \right]$$

$$\lambda_{br} = \frac{C_b X_1}{(F_y - 10)} \sqrt{1 + \sqrt{1 + \frac{4X_2(F_y - 10)^2}{C_b^2 X_1^2}}}$$

$\lambda_b \leq \lambda_{br}$ — Yes → $M_u = M_p - (M_p - M_r)\left(\dfrac{\lambda_b - \lambda_{bp}}{\lambda_{br} - \lambda_{bp}} \right)$

No ↓

$$M_u = \frac{S_x C_b X_1}{\lambda_b} \sqrt{1 + \frac{X_2}{\lambda_b^2}}$$

Flowchart 9.2

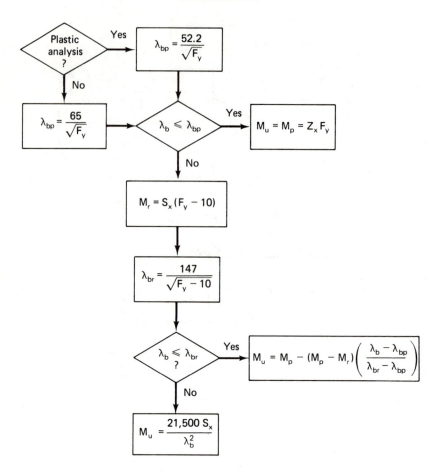

Flowchart 9.3

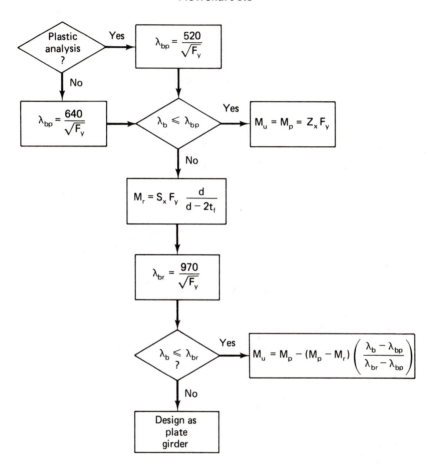

Solution

Case 1 : *Braced Beam*

$$\text{dead load} = 1.20 \text{ kips/ft}$$

$$\text{live load} = 0.74 \text{ kips/ft}$$

Use plastic analysis for a braced continuous beam.

$$M_x = \left(\frac{wL}{2} - \frac{M_p}{L}\right) x - \frac{wx^2}{2}$$

$M_{\max}$ occurs when the derivative of M_x goes to zero.

$$\frac{dM_x}{dx} = \frac{wL}{2} - \frac{M_p}{L} - wx = 0 \qquad \bar{x} = \frac{L}{2} - \frac{M_p}{wL}$$

$$M_{\max} \equiv M_p = \frac{wL^2}{2}\left(\frac{1}{2} - \frac{M_p}{wL^2}\right)^2$$

Solving,

$$\frac{2M_p}{wL^2} - \left(\frac{1}{2} - \frac{M_p}{wL^2}\right)^2 = 0$$

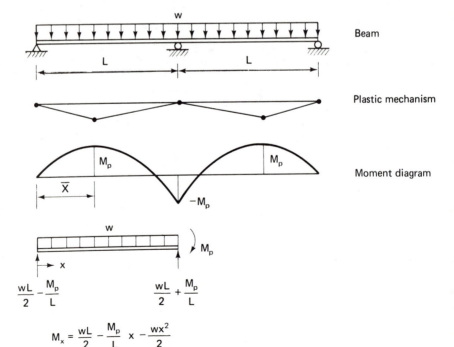

Beam

Plastic mechanism

Moment diagram

$$M_x = \frac{wL}{2} - \frac{M_p}{L} \ x - \frac{wx^2}{2}$$

we get

$$\frac{M_p}{wL^2} = \frac{3}{2} \pm \sqrt{2}$$

The smallest value controls; therefore,

$$M_p = wL^2\left(\frac{3}{2} - \sqrt{2}\right) = \frac{wL^2}{11.657}$$

LRFD requirement:

$$\phi M_p \geq \gamma_A(\gamma_D w_D + \gamma_L w_L)\frac{L^2}{11.657}$$

$$M_p = Z_x F_y$$

Therefore,

$$Z_x \geq \frac{1}{\phi F_y}\left[\gamma_A(\gamma_D w_D + \gamma_L w_L)\frac{L^2}{11.657}\right]$$

for $\phi = 0.86$, $F_y = 36$ ksi,

$$\gamma_A = 1.1 \qquad \gamma_D = 1.1 \qquad \gamma_L = 1.4$$
$$w_D = 1.20 \text{ kips/ft} \qquad w_L = 0.74 \text{ kip/ft} \qquad L = 30 \text{ ft}$$

$$Z_x \geq \frac{1}{0.86 \times 36}\left[1.1(1.1 \times 1.20 + 1.4 \times 0.74) \times \frac{30^2}{11.657}\right] \times 12 = 77.6 \text{ in.}^3$$

Try W 18 × 40:

$$Z_x = 78.4 \text{ in.}^3 \qquad \text{(AISCM)}$$

Check if this section is acceptable for plastic design. For W 18 × 40:

$$r_y = 1.27 \text{ in.} \qquad \frac{b_f}{2t_f} = 5.73 \qquad \frac{d}{t_w} = 56.8$$

From Flowchart 9.1, limit state LTB, the moment is nearly uniform near the plastic hinge in the span. Therefore,

$$\lambda_{bp} = \frac{240}{\sqrt{F_y}} = \frac{240}{\sqrt{36}} = 40.0$$

The unbraced length = joist spacing, $L_b = 2$ ft = 24 in.:

$$\lambda_b = \frac{L_b}{r_y} = \frac{24}{1.27} = 18.9 < 40.0 \qquad \text{OK}$$

From Flowchart 9.2, limit state FLB:

$$\lambda_{bp} = \frac{52.2}{\sqrt{F_y}} = \frac{52.2}{\sqrt{36}} = 8.7$$

$$\lambda_b = \frac{b_f}{2t_f} = 5.73 < 8.7 \qquad \text{OK}$$

From Flowchart 9.3, limit state **WLB**:

$$\lambda_{bp} = \frac{520}{\sqrt{F_y}} = \frac{520}{\sqrt{36}} = 86.7$$

$$\lambda_b = \frac{d}{t_w} = 56.8 < 86.7 \qquad \text{OK}$$

Therefore, use W 18 × 40.

Case 2: *Construction loading*

$$\text{dead load} = 0.04 \text{ kip/ft} \qquad \text{(beam weight only)}$$

$$\text{construction load, } P_c = 5 \text{ kips at the center of each span}$$

Factored loads:

$$w = \gamma_A (\gamma_D w_D) = 1.1 \times 1.1 \times 0.04 = 0.0484 \text{ kip/ft} \cong 0.05 \text{ kip/ft}$$

$$P = \gamma_A (\gamma_C P_C) = 1.1 \times 1.4 \times 5 = 7.70 \text{ kip}$$

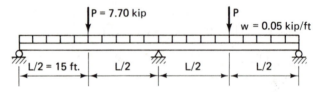

From the AISCM, the negative elastic moment under the center support is

$$M_S = -\frac{wL^2}{8} - \frac{3PL}{16} = \left(-\frac{0.05 \times 30^2}{8} - \frac{3 \times 7.7 \times 30}{16}\right) \times 12 = -587 \text{ kip-in.}$$

The maximum positive moment occurs under each concentrated load:

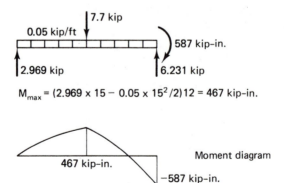

$$M_{max} = (2.969 \times 15 - 0.05 \times 15^2/2)12 = 467 \text{ kip-in.}$$

The unbraced length is $L_b = 30$ ft $= 360$ in. Since no lateral bracing is provided between the beam supports for the case of construction loading, $C_b = 1.0$ will be used. This is a conservative assumption.

Check the adequacy of W 18 × 40. Only limit state LTB needs to be considered again, since FLB and WLB were previously found to be OK.

From AISCM, for W 18 × 40:

$$r_y = 1.27 \text{ in.} \qquad A = 11.8 \text{ in.}^2$$
$$S_x = 68.4 \text{ in.}^3 \qquad d = 17.90 \text{ in.}$$
$$J = 0.809 \text{ in.}^4 \qquad t_f = 0.525 \text{ in.}$$
$$C_w = 1440 \text{ in.}^6$$

Following Flowchart 9.1:

$$\lambda_b = \frac{L_b}{r_y} = \frac{360}{1.27} = 283.5$$

$$\lambda_{bp} = \frac{240}{\sqrt{F_y}} = \frac{240}{\sqrt{36}} = 40.0 < \lambda_b$$

$$M_r = S_x(F_y - 10) = 68.4(36 - 10) = 1778 \text{ kip-in.}$$

$$X_1 = \frac{56,500\sqrt{JA}}{S_x} = \frac{56,500\sqrt{0.809 \times 11.8}}{68.4} = 2552 \text{ ksi}$$

$$X_2 = 6.41 \left[\frac{A(d - t_f)^2}{J}\right] = \frac{6.41 \times 11.8(17.90 - 0.525)^2}{0.809} = 28,225$$

$$\lambda_{br} = \frac{C_b X_1}{F_y - 10} \sqrt{1 + \sqrt{1 + \frac{4 X_2 (F_y - 10)^2}{C_b^2 X_1^2}}}$$

$$= \frac{1 \times 2552}{36 - 10} \sqrt{1 + \sqrt{1 + \frac{4 \times 28,225(36 - 10)^2}{1^2 \times 2552^2}}} = 210 < 283.5$$

$$M_u = \frac{S_x C_b X_1}{\lambda_b} \sqrt{1 + \frac{X_2}{\lambda_b^2}} = \frac{68.4 \times 1 \times 2552}{283.5} \sqrt{1 + \frac{28,225}{(283.5)^2}} = 716 \text{ kip-in.}$$

$$\phi M_u = 0.86 \times 716 = 616 \text{ kip-in.} > 587 \text{ kip-in.}$$

Therefore, W 18 × 40 is OK for construction loads also. Use W 18 × 40.

Example 9.3

Design a pinned-end column in a braced frame to support the following loads:

$$\text{mean dead load} = 300 \text{ kips}$$
$$\text{mean wind load} = 450 \text{ kips}$$

The length of the column is 18 ft 0 in. = 216 in. The material is A36 steel, $F_y = 36$ ksi.

LRFD Criteria (Sec. 2.3.2, Ref. 9.1):

$$\gamma_A(\gamma_D P_D + \gamma_W P_W) \le \phi_c P_u$$
$$\gamma_A = 1.1 \qquad \gamma_D = 1.1 \qquad \gamma_W = 1.6$$

Design criteria are given in Flowchart 9.4.

Flowchart 9.4

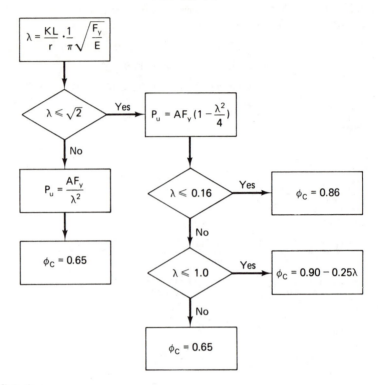

Solution

The factored axial load is

$$\gamma_A(\gamma_D P_D + \gamma_W P_W) = 1.1(1.1 \times 300 + 1.6 \times 450) = 1155 \text{ kips}$$

Try W 14 × 145:

$$A = 42.7 \text{ in.}^2 \qquad r_y = 3.98 \text{ in.}$$

$$\lambda = \frac{KL}{r_y} \frac{1}{\pi} \sqrt{\frac{F_y}{E}} = \frac{1.0 \times 216}{3.98\pi} \sqrt{\frac{36}{29,000}} = 0.609$$

$$\lambda < \sqrt{2} \qquad 0.16 < \lambda < 1.0$$

$$P_u = AF_y\left(1 - \frac{\lambda^2}{4}\right) = 42.7 \times 36\left(1 - \frac{0.609^2}{4}\right) = 1395 \text{ kips}$$

$$\phi = 0.90 - 0.25\lambda = 0.90 - 0.25 \times 0.609 = 0.748$$

$$\phi P_u = 0.748 \times 1395 = 1043 \text{ kips} < 1155 \qquad \text{NG}$$

Try W 14 × 159:

$$A = 46.7 \text{ in.}^2 \qquad r_y = 4.00 \text{ in.}$$

$$\lambda = 0.606 \qquad P_u = 1527 \text{ kips} \qquad \phi = 0.749$$

$$\phi P_u = 1144 \text{ kips} \approx 1155 \quad (\text{within } 1\%) \quad \text{OK}$$

Use W 14 × 159.

Example 9.4

Select a column for a 24-ft story height to support the following forces:

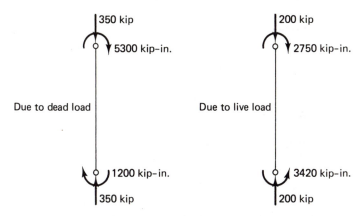

|350 kip
5300 kip-in.

Due to dead load

1200 kip-in.
350 kip

|200 kip
2750 kip-in.

Due to live load

3420 kip-in.
200 kip

Bending is about the x axis only. The column is prevented from sidesway in both the x and the y directions. Lateral support against movement in the y direction, and against twist, is provided at the ends and at the center. Use A36 steel:

$$F_y = 36 \text{ ksi} \qquad E = 29,000 \text{ ksi}$$

LRFD Design requirements (Ref. 9.1); The load factors are the same as in Example 9.2.

Step 1: *Check the Axial Capacity*

The factored design axial load $\leq \phi_c P_u$. P_u and ϕ_c are determined from Flowchart 9.4.

Step 2: *Check the Moment Capacity—Strength*

a. *Compact section:* when

$$\frac{b_f}{2t_f} \leq \frac{65}{\sqrt{F_y}}$$

and

$$\frac{d}{t} \leq \begin{cases} \dfrac{640}{\sqrt{F_y}}\left(1 - 2.75\dfrac{P}{\phi_b P_y}\right) & \text{for} \quad \dfrac{P}{\phi_b P_y} \leq 0.125 \\[2ex] \dfrac{152}{\sqrt{F_y}}\left(2.89 - \dfrac{P}{\phi_b P_y}\right) & \text{for} \quad \dfrac{P}{\phi_b P_y} > 0.125 \end{cases}$$

$$\frac{P}{\phi_b P_y} + \frac{M_2}{1.18\phi_b M_p} \leq 1.0 \quad \text{and} \quad M_2 \leq \phi_b M_p$$

where

$$P = \text{factored axial load}$$
$$P_y = AF_y, \text{ yield load}$$
$$\phi_b = 0.86, \text{ resistance factor}$$
$$M_2 = \text{maximum factored end moment}$$
$$M_p = Z_x F_y, \text{ plastic moment}$$

b. *Noncompact section:* when $b_f/2t_f$ and/or d/t exceed the limits given above, the following interaction equation must be checked:

$$\frac{P}{\phi_b P_y} + \frac{M_2}{\phi_b M_u} \le 1.0$$

where M_u is determined from Flowchart 9.2 and/or 9.3, as appropriate.

Step 3: *Check the Moment Capacity—Instability*

The following interaction equation must be satisfied:

$$\frac{P}{\phi_b P_u} + \frac{C_m M_2}{\phi_b M_u (1 - P/\phi_b P_{Ex})} \le 1.0$$

where

P_u = axial capacity, as determined from Flowchart 9.4

M_u = moment capacity, smallest of values from Flowchart 9.1, 9.2, or 9.3, respectively; when lateral torsional buckling limit state is investigated (Flowchart 9.1), use $C_b = 1.0$ when sidesway is prevented, and C_b from Flowchart 9.1 if sidesway is permitted

$$C_m = \begin{cases} 0.85 & \text{when sidesway is permitted} \\ 0.6 - 0.4(M_1/M_2) \le 0.4 & \text{when sidesway is prevented} \end{cases}$$

M_1 is the numerically smaller and M_2 the larger end moment. The ratio M_1/M_2 is positive when bending causes reverse curvature.

$$P_{Ex} = \frac{AF_y}{\lambda_x^2}$$

$$\lambda_x = \frac{K_x L}{r_x} \frac{1}{\pi} \sqrt{\frac{F_y}{E}}$$

Solution

Factored design loads:
Axial load:

$$P = \gamma_A(\gamma_D P_D + \gamma_L P_L) = 1.1(1.1 \times 350 + 1.4 \times 200) = 731.5 \text{ kips}$$

End moments:

$$M_{\text{top}} = 1.1(1.1 \times 5300 + 1.4 \times 2750) = 10,648 \text{ kip-in.}$$
$$M_{\text{bottom}} = 1.1(1.1 \times 1200 - 1.4 \times 3420) = -3815 \text{ kip-in.}$$

$$M_2 = 10,648 \text{ in.-kip}$$

$$\frac{M_1}{M_2} = -\frac{3815}{10,650} = -0.358$$

Sidesway is prevented:

$$C_m = 0.6 - (0.4)(-0.358) = 0.743$$

Design lengths:

$$K_x L_x = 1.00 \times 24 \times 12 = 288 \text{ in.}$$
$$K_y L_y = 1.00 \times 12 \times 12 = 144 \text{ in.}$$
$$\text{unbraced length} = 144 \text{ in.}$$

Preliminary selection guidelines:

$$\frac{P}{F_y} = \frac{732}{36} \cong 20 \text{ in.}^2$$

$$\frac{M}{F_y} = \frac{10,650}{36} \cong 300 \text{ in.}^3$$

Try a W 30 × 173:

$$A = 50.8 \text{ in.}^2 \qquad P_y = F_y A = 36 \times 50.8 = 1828.8 \text{ kips}$$

$$S_x = 539 \text{ in.}^3 \qquad M_r = S_x(F_y - 10) = 539(36 - 10) = 14,014 \text{ kip-in.}$$

$$r_x = 12.7 \text{ in.} \qquad \frac{K_x L_x}{r_x} = \frac{288}{12.7} = 22.68 \qquad \lambda_x = \frac{22.68}{\pi}\sqrt{\frac{F_y}{E}} = 0.254$$

$$r_y = 3.43 \text{ in.} \qquad \frac{K_y L_y}{r_y} = \frac{144}{3.43} = 41.98 \qquad \lambda_y = \frac{41.98}{\pi}\sqrt{\frac{F_y}{E}} = 0.471$$

$$\frac{b_f}{2t_f} = 7.04 \qquad \frac{65}{\sqrt{F_y}} = \frac{65}{\sqrt{36}} = 10.83 > 7.04$$

$$\frac{d}{t} = 46.5$$

$$\frac{P}{P_y} = \frac{731.5}{1828.8} = 0.400 \qquad \phi_b P_y = 0.86 \times 0.400 = 0.344 > 0.125$$

$$\frac{152}{\sqrt{F_y}}\left(2.89 - \frac{P}{\phi_b P_y}\right) = 64.50 > 46.5$$

The section is compact.

$$Z_x = 605 \text{ in.}^3 \qquad M_p = Z_x F_y = 605 \times 36 = 21,780 \text{ kip-in.}$$

Following Flowchart 9.4:

$$\lambda_y > \lambda_x \qquad \lambda = \lambda_y = 0.471 < \sqrt{2}$$

$$P_u = AF_y\left(1 - \frac{\lambda^2}{4}\right) = 1828.8\left(1 - \frac{0.471^2}{4}\right) = 1727 \text{ kips}$$

$$0.16 \leq \lambda \leq 1.0 \qquad \phi_c = 0.90 - 0.25\lambda = 0.782$$

Check the axial capacity:

$$\phi_c P_u = 0.782 \times 1727 = 1351 \text{ kips} > 731.5 \text{ kips} \qquad \text{OK}$$

Check the strength:

$$\frac{P}{\phi_b P_y} + \frac{M_2}{1.18\phi_b M_p} = \frac{731.5}{0.86 \times 1828.8} + \frac{10{,}648}{1.18 \times 0.86 \times 21{,}780} = 0.947 < 1.0$$

OK

Check the stability:
Limit-state lateral-torsional buckling (Flowchart 9.1):

$$\lambda_b = \frac{L_b}{r_y} = \frac{L_y}{r_y} = 41.98$$

Moment gradient:

$$\frac{M_1}{M_2} = -0.358 > -0.5$$

$$\lambda_{bp} = \frac{390}{\sqrt{F_y}} = 65 > 41.98$$

Therefore,

$$M_u = M_p = 21{,}780 \text{ kip-in.}$$

$$P_{Ex} = \frac{AF_y}{\lambda_x^2} = \frac{1828.8}{0.254^2} = 28{,}346 \text{ kips}$$

$$\frac{P}{\phi_b P_u} + \frac{C_m M_2}{\phi_b M_u \left(1 - \dfrac{P}{\phi_b P_{Ex}}\right)} = \frac{731.5}{0.86 \times 1727} + \frac{0.743 \times 10{,}648}{0.86 \times 21{,}780\left(1 - \dfrac{731.5}{0.86 \times 28{,}346}\right)}$$

$$= 0.928$$

$$0.928 < 1.00 \quad \text{OK}$$

Use a W 30 × 173.

Example 9.5

Design a plate girder for the loads shown.

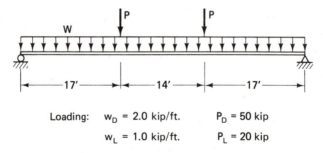

Loading: w_D = 2.0 kip/ft. P_D = 50 kip
           w_L = 1.0 kip/ft. P_L = 20 kip

Factored loads:

$$w = \gamma_A(\gamma_D w_D + \gamma_L w_L) = 1.1(1.1 \times 2 + 1.4 \times 1) = 3.960 \text{ kips/ft}$$

$$P = \gamma_A(\gamma_D P_D + \gamma_L P_L) = 1.1(1.1 \times 50 + 1.4 \times 20) = 91.300 \text{ kips}$$

The material is A36 steel:

$$F_y = 36 \text{ ksi} \qquad E = 29{,}000 \text{ ksi}$$

Use an E70 electrode, $F_{Exx} = 70$ ksi. The maximum depth is 72 in.
  LRFD Design requirements (Ref. 9.1):

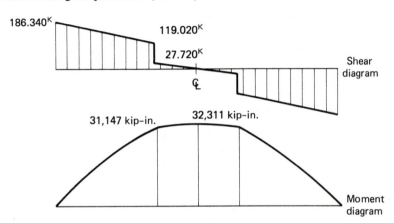

Step 1: *Moment Capacity*

Geometry:

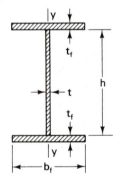

$$r_T = \frac{b_f}{\sqrt{12}} \sqrt{\frac{1}{1 + \frac{1}{6} \cdot \frac{ht}{b_f t_f}}}$$

Radius of gyration of the flange
plus 1/6 of the web about the
y - axis.

$$A_w = ht; \quad A_f = b_f t_f$$

a : Stiffener spacing

Geometric limits:

$$\frac{h}{t} \le \left(\frac{h}{t}\right)_{\max}$$

where

$$\left(\frac{h}{t}\right)_{\max} = \frac{2000}{\sqrt{F_y}} \qquad \text{if } \frac{a}{h} \le 1.5$$

and

$$\left(\frac{h}{t}\right)_{\max} = \frac{14{,}000}{\sqrt{F_y(F_y + 16.5)}} \qquad \text{if } \frac{a}{h} > 1.5$$

$$\frac{a}{h} \le 3.0 \qquad \text{and} \qquad \frac{a}{h} \le \left(\frac{260}{h/t}\right)^2$$

whichever is smaller.

Design criterion:

$$M \leq \phi M_u$$

$$\phi = 0.86$$

Determine $M_u = S_x R_{PG} F_{cr}$

```
        ┌──────────────┐   Yes   ┌──────────────┐
        │ h    970     │ ──────→ │  Design as a │
        │ ─ ≤ ────     │         │ beam, Example 2│
        │ t    √Fy     │         └──────────────┘
        └──────────────┘
              │ No
              ▼
```

Limit state LTB	Limit state FLB
$\lambda_b = \dfrac{L_b}{r_T}$	$\lambda_b = \dfrac{b_f}{2t_f}$
$\lambda_{bp} = \dfrac{146}{\sqrt{F_y}}$	$\lambda_{bp} = \dfrac{52.2}{\sqrt{F_y}}$
$\lambda_{br} = \dfrac{757\sqrt{C_b}}{\sqrt{F_y}}$	$\lambda_{br} = \dfrac{149}{\sqrt{F_y}}$
$C_{PG} = 286{,}000\, C_b$	$C_{PG} = 11{,}140$

```
         ┌──────────────┐  Yes  ┌──────────────┐
         │ λb ≤ λbp     │ ────→ │  Fcr = Fy    │ ──────┐
         └──────────────┘       └──────────────┘       │
              │ No                                       │
              ▼                                          │
         ┌──────────────┐  Yes                           │
         │ λb = λbr     │ ────→                          │
         └──────────────┘                                │
              │ No                                       │
              ▼                                          │
```

$F_{cr} = F_y \left[1 - \dfrac{0.5(\lambda_b - \lambda_{bp})}{\lambda_{br} - \lambda_{bp}} \right]$

$F_{cr} = \dfrac{C_{PG}}{\lambda_b^2}$

$R_{PG} = 1 - \dfrac{A_w / A_f}{2000} \left[\dfrac{h}{t} - \dfrac{970}{\sqrt{F_{cr}}} \right]$

$$M_u = S_x R_{PG} F_{cr}$$

Step 2: *Shear Capacity*

Design criterion:

$$V \leq \phi V_u$$

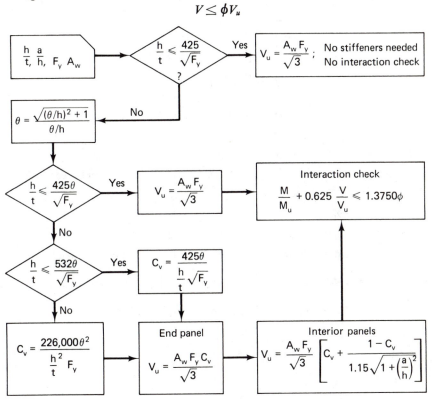

Step 3: *Stiffener Design*

No stiffeners are needed if

$$\frac{h}{t} \leq \frac{425}{\sqrt{F_y}} \quad \text{and} \quad V \leq \frac{\phi A_w F_y C_v}{\sqrt{3}}$$

where $\phi = 0.86$ and C_v is determined for $\theta = 1$.

Stiffener requirements: Moment of inertia about an axis in the web center $\geq$ $at^3 j$, where

$$j = \frac{2.5}{(a/h)^2} - 2 \geq 0.5$$

$$\text{stiffener area} \geq \frac{(F_y)_{\text{web}}}{(F_y)_{\text{stiffener}}} \left[0.15 \, Dht \, (1 - C_v) \frac{V}{\phi V_u} - 18t^2 \right]$$

where

$$
\begin{aligned}
D &= 1 && \text{for stiffeners in pairs} \\
D &= 1.8 && \text{for single angle stiffeners} \\
D &= 2.4 && \text{for single plate stiffeners}
\end{aligned}
$$

Bearing stiffeners are designed as columns (Example 9.3), with an effective length $KL = 0.75h$ and a cross section comprised of the two stiffeners and a strip of the web as shown:

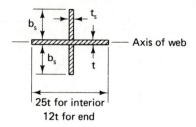

No bearing stiffeners are required when

$$\text{reaction or} \atop \text{concentrated force} \begin{cases} \leq \phi_{bs}t(N + 2k)F_y & \text{for interior concentrated loads} \\ \leq \phi_{bs}t(N + k)F_y & \text{for end reaction} \end{cases}$$

$\phi_{bs} = 0.92$

N = length of bearing $\geq k$ at end reactions

k = distance from outer face of flange to web toe of fillet weld

Step 4: *Fillet-Weld Design*

weld stress on effective fillet-weld area $\leq \phi F_w$

$$\phi = 0.80 \qquad F_w = 0.6F_{Exx}$$

$$F_{Exx} = \text{tensile strength of weld}$$

Solution

Trial section:

$$\text{Web:} \qquad \tfrac{5}{16} \text{ in.} \times 70 \text{ in.}$$

$$\text{Flanges:} \qquad \tfrac{7}{8} \text{ in.} \times 18.5 \text{ in.}$$

$$\text{total depth } 71.75 \text{ in.} < 72 \text{ in.} \qquad \text{OK}$$

$t = 0.3125 \text{ in.} \qquad \dfrac{h}{t} = 224 \qquad A_w = ht = 21.875 \text{ in.}^2$

$h = 70 \text{ in.}$

$b_f = 18.5 \text{ in.} \qquad \dfrac{b_f}{2t_f} = 10.571 \qquad A_f = b_f t_f = 16.1875 \text{ in.}^2$

$t_f = 0.875 \text{ in.}$

$r_T = \dfrac{b_f}{\sqrt{12}} \sqrt{\dfrac{1}{1 + A_w/6A_f}} = 4.825 \text{ in.}$

$I_x = \dfrac{b_f t_f^3}{12} \times 2 + 2A_f \left(\dfrac{h + t_f/2}{2}\right)^2 + \dfrac{th^3}{12} = 49{,}901 \text{ in.}^4$

$$S_x = \frac{2I_x}{h + 2t_f} = 1368 \text{ in.}^3$$

$$\left(\frac{h}{t}\right)_{\max} = \frac{2000}{\sqrt{F_y}} = 333 \quad \text{or} \quad \left(\frac{h}{t}\right)_{\max} = \frac{14{,}000}{\sqrt{F_y(F_y + 16.5)}} = 322 > 224 \qquad \text{OK}$$

Check the flexure:

$$\frac{970}{\sqrt{F_y}} = 162 < 224 \qquad \text{plate girder design applies}$$

Limit state LTB: Lateral support at the ends and under the concentrated loads:
Outside span:

$$\frac{M_1}{M_2} = 0$$

$$C_b = 1.75 + 1.05\left(\frac{M_1}{M_2}\right) + 0.3\left(\frac{M_1}{M_2}\right)^2 = 1.75$$

$$L_b = 17 \text{ ft} = 204 \text{ in.}$$

$$\frac{L_b}{r_T} = \frac{204}{4.825} = 42.28 = \lambda_b$$

$$\lambda_{bp} = \frac{146}{\sqrt{F_y}} = 24.33 < 42.28$$

$$\lambda_{br} = \frac{757\sqrt{C_b}}{\sqrt{F_y}} = 166.90 > 42.28$$

$$F_{cr} = F_y\left[1 - \frac{0.5(\lambda_b - \lambda_{bp})}{\lambda_{br} - \lambda_{bp}}\right] = 33.73 \text{ ksi}$$

$$R_{pG} = 1 - \frac{A_w}{2000A_f}\left(\frac{h}{t} - \frac{970}{\sqrt{F_{cr}}}\right) = 0.961$$

$$M_u = S_x R_{pG} F_{cr} = 44{,}371 \text{ kip-in.}$$

Center span:

$$\frac{M_1}{M_2} = -\frac{31{,}147}{32{,}311} = -0.964$$

$$C_b = 1.75 + 1.05(-0.964) + 0.3(0.964)^2 = 1.017$$

$$L_b = 14 \text{ ft} = 168 \text{ in.}$$

$$\frac{L_b}{r_T} = \frac{168}{4.825} = 34.82$$

$$\lambda_{bp} = 24.33 < 34.82$$

$$\lambda_{br} = 127.21 > 34.82$$

$$F_{cr} = F_y\left[1 - \frac{0.5(34.82 - 24.33)}{127.21 - 24.33}\right] = 34.165 \text{ ksi}$$

$$R_{pG} = 1 - \frac{21.875}{2000 \times 16.187}\left(224 - \frac{970}{\sqrt{34.165}}\right) = 0.961$$

$$M_u = 1368 \times 0.961 \times 34.165 = 44{,}905 \text{ kip-in.}$$

Limit state FLB:

$$\lambda_b = \frac{b_f}{2t_f} = 10.571$$

$$\lambda_{bp} = \frac{52.2}{\sqrt{F_y}} = 8.700 < 10.571$$

$$\lambda_{br} = \frac{149}{\sqrt{F_y}} = 24.833 > 10.571$$

$$F_{cr} = F_y\left[1 - \frac{0.5(\lambda_b - \lambda_{bp})}{\lambda_{br} - \lambda_b}\right] = 33.912 \text{ ksi}$$

$$R_{pG} = 0.961$$

$$M_u = 44,591 \text{ kip-in.}$$

Least value:

$$M_u = 44,371 \text{ kip-in.}$$

$$\phi M_u = 0.86 M_u = 38,159 \text{ kip-in.} > 32,311 \text{ kip-in.} \qquad \text{OK}$$

So the flexure is OK.

Check the shear:

End panel: Assume that

$$a = 42 \text{ in.} \qquad \frac{a}{h} = \frac{42}{70} = 0.600$$

$$\frac{h}{t} = 224$$

$$\frac{425}{\sqrt{F_y}} = 70.83 < 224$$

$$\theta = \frac{\sqrt{(a/h)^2 + 1}}{a/h} = 1.944$$

$$\frac{425\theta}{\sqrt{F_y}} = 137.675 < 224$$

$$\frac{532\theta}{\sqrt{F_y}} = 172.337 < 224$$

$$C_v = \frac{226,000\theta^2}{(h/t)^2 F_y} = 0.473$$

$$V_u = \frac{A_w F_y C_v}{\sqrt{3}} = 214.90 \text{ kips} > 186.34 \text{ kips} \qquad \text{OK}$$

Interior panel: Assume that

$$a = 81 \text{ in.} \qquad \frac{a}{h} = 1.157 < 3 \qquad \text{OK}$$

$$\left(\frac{260}{h/t}\right)^2 = 1.347 > 1.157 \qquad \text{OK}$$

$$\theta = 1.322$$

$$\frac{532\theta}{\sqrt{F_y}} = 117.195 < 224$$

$$C_v = \frac{226,000\theta^2}{(h/t)^2 F_y} = 0.219$$

$$V_u = \frac{A_w F_y}{\sqrt{3}}\left[C_v + \frac{1 - C_v}{1.15\sqrt{1 + (a/h)^2}}\right] = 301.48 \text{ kips} > 186.34 \text{ kips}\qquad \text{OK}$$

Stiffener spacing:

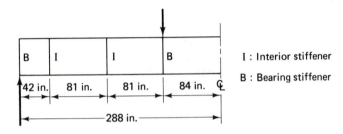

B : Bearing stiffener

I : Interior stiffener

Interaction check: under load:

$$M = 31,147 \text{ kip-in.}\qquad V = 119.02 \text{ kips}$$

$$M_u = 44,371 \text{ kip-in.}\qquad V_u = 301.48 \text{ kips}$$

$$\frac{M}{M_u} + 0.625\frac{V}{V_u} = 0.949$$

$$1.375\phi = 1.375 \times 0.86 = 1.183 > 0.949\qquad \text{OK}$$

Use a $\frac{5}{16} \times 70$ in. web and $\frac{7}{8} \times 18.5$ in. flanges.
Interior stiffener design: single plate stiffener:

$$\frac{b_s}{t_s} \leq \frac{95}{\sqrt{F_y}} = 15.83$$

Local buckling;
(AISCS 1.9.1)

$$I_s = \frac{t_s(b_s + t/2)^3}{3}$$

$$I_s \geq at^3j$$

$$a = 81 \text{ in}\qquad t = 0.3125 \text{ in.}$$

$$j = \frac{2.5}{(a/h)^2} - 2 = \frac{2.5}{(1.157)^2} - 2 = -0.13 < 0.5\qquad \therefore\quad j = 0.5$$

$$\frac{t_s(b_s + t/2)^3}{3} \geq 81 \times 0.3125^3 \times 0.5 = 1.236 \text{ in.}^4$$

$$A_s = t_s b_s \geq 0.15 \, Dht(1 - C_v)\frac{V}{\phi V_u} - 18t^2$$

For a single plate stiffener, $D = 2.4$:

$$C_v = 0.219 \qquad V = 186.340 \text{ kips} \qquad \phi = 0.86 \qquad V_u = 301.483 \text{ kips}$$

$$t_s b_s \geq 0.15 \times 2.4 \times 70 \times 0.3125(1 - 0.219)\frac{186.340}{0.86 \times 301.483} - 18 \times 0.3125^2$$

$$= 2.662 \text{ in.}^2$$

Try a $\frac{1}{2} \times 6$ in. plate:

$$\frac{b_s}{t_s} = 12 < 15.83 \qquad \text{OK}$$

$$I_s = 38.886 \text{ in.}^4 > 1.236 \text{ in.}^4 \qquad \text{OK}$$

$$A_s = 3 \text{ in.}^2 > 2.662 \text{ in.}^2 \qquad \text{OK}$$

Use $\frac{1}{2} \times 6$ in. single plate interior stiffeners.
Bearing stiffener design:
At the reaction end (critical):

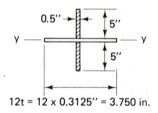

12t = 12 x 0.3125" = 3.750 in.

Try $\frac{1}{2} \times 5$ in. plates.

$$I_y = \frac{0.5(5 \times 2 + 0.3125)^3}{12} = 45.696 \text{ in.}^4$$

$$A = 2 \times 5 \times 0.5 + 3.75 \times 0.3125 = 6.172 \text{ in.}^2$$

$$r = \sqrt{\frac{I_y}{A}} = 2.721 \text{ in.}$$

$$\frac{KL}{r} = \frac{0.75h}{r} = \frac{0.75 \times 70}{2.721} = 19.294 \left.\vphantom{\begin{array}{c} \\ \\ \\ \\ \\ \\ \end{array}}\right\}$$

$$\lambda = \frac{KL}{r}\frac{1}{\pi}\sqrt{\frac{F_y}{E}} = 0.216 \qquad \text{Example 9.3}$$

$$P_{cr} = AF_y\left(1 - \frac{\lambda^2}{4}\right) = 219.591 \text{ kips}$$

$$\phi = 0.9 - 0.25\lambda = 0.846$$

$$\phi P_{cr} = 185.753 \text{ kips} \cong 186.340 \text{ kips within } 1\%$$

Use $\frac{1}{2} \times 5$ in. plates for bearing stiffeners.

Design of fillet weld between flange and web:

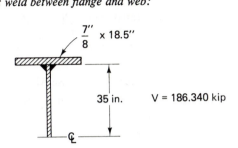

Shear flow across weld : $F_s = \dfrac{V}{I_x} \cdot (\dfrac{7}{8} \times 18.5)\ (35 + \dfrac{1}{2} \times \dfrac{7}{8})$

$I_x = 49{,}901\ \text{in.}^4$

$F_s = 2.142\ \text{kip/in.}$

Weld throat:

$$\frac{D}{\sqrt{2}} \qquad \text{where } D = \text{weld size}$$

Weld stress:

$$f_v = \frac{2.142\sqrt{2}}{D}$$

Weld capacity:

$$\phi F_x = 0.80 \times 0.6 F_{Exx} = 0.80 \times 0.6 \times 70 = 33.60\ \text{ksi}$$

$$\frac{2.142\sqrt{2}}{D} = 33.60 \qquad D = 0.09\ \text{in.}$$

Minimum required weld size (AISCS, Sec. 1.17): $D = \frac{5}{16}$ in. Use $\frac{5}{16}$-in. fillet weld.

Check if bearing stiffener is required under concentrated load:

$$91.3\ \text{kips} \le \phi_{bs} t (N + 2k) F_y$$

$\phi_{bs} = 0.92 \qquad t = 0.3125\ \text{in.} \qquad F_y = 36\ \text{ksi} \qquad k = \frac{7}{8} + \frac{5}{16} = 1.19\ \text{in.}$

Assume that the bearing length is 4 in.:

$$\phi_{bs} t (N + 2k) F_y = 65.98\ \text{kips} < 91.3\ \text{kips}$$

A bearing stiffener is needed.

PROBLEMS

9.1. Same as Example 9.1, but the span is 50 ft.

9.2. Design a fixed-ended wide-flange beam ($F_y = 50$ ksi) of 40-ft length by plastic design using the LRFD criteria given in Example 9.2. The beam supports a dead load of 80 lb/ft² and an office live load. The beam spacing is 20 ft. The beam is laterally supported along its whole length.

9.3. Design a simply supported wide-flange beam using LRFD ($F_y = 36$ ksi) which is subjected to a central concentrated load of 30 kips (15 kips from dead load and 15 kips from live load) in addition to its own weight. The span of the beam is 60 ft, and lateral support exists only at the supports and under the concentrated load.

9.4. Design the same beam as in Problem 9.3, but use AISCS.

9.5. Design the column in Example 9.3 assuming that the 450-kip force derives from the live load, using LRFD.

9.6. Same as Problem 9.5, using AISCS.

9.7. Design the column in Example 9.4 using LRFD, but with the 1200-kip-in. bottom dead-load moment reversed in direction.

9.8. Design a plate girder, using LRFD, using double the loads of Example 9.5. There is no imposed depth limit. Use steel with $F_y = 50$ ksi.

REFERENCES

9.1. "Proposed Criteria for Load and Resistance Factor Design of Steel Building Structures." AISI *Bulletin 27.* (1978).

9.2. "Building Code Requirements for Minimum Design Loads in Buildings and Other Structures." ANSI *A58.1* (1972).

9.3. T. V. GALAMBOS AND M. K. RAVINDRA. "Load and Resistance Factor Design for Steel," *ASCE Journal of the Structural Division*, Vol. ST9 (Sept. 1978).

9.4. J. A. YURA, T. V. GALAMBOS, AND M. K. RAVINDRA. "The Bending Resistance of Steel Beams." *ASCE Journal of the Structural Division*, Vol. ST9 (Sept. 1978).

9.5. P. B. COOPER, T. V. GALAMBOS, AND M. K. RAVINDRA. "LRFD Criteria for Plate Girders." *ASCE Journal of the Structural Division.* Vol. ST9 (Sept. 1978).

9.6. C. W. PINKHAM AND W. C. HANSELL. "An Introduction to Load and Resistance Factor Design for Steel Buildings." *AISC Engineering Journal*, Vol. 15, No. 1 (1978).

9.7. T. V. GALAMBOS AND M. K. RAVINDRA. "Proposed Criteria for Load and Resistance Factor Design." *AISC Engineering Journal*, Vol. 15, No. 1 (1978).

9.8. K. B. WIESNER. "LRFD Design Office Study." *AISC Engineering Journal*, Vol. 15, No. 1 (1978).

10

SPECIAL TOPICS
IN BEAM DESIGN

10.1 INTRODUCTION

Chapters 3, 7, and 8 included beam and plate girder design problems for which specification coverage is adequate, including the usual problems arising from lack of lateral support of rolled shapes having an axis of symmetry in the plane of the loads. Chapter 9 applied load and resistance factor design to similar problems. We now consider special problems that arise in beam design as a result of lack of lateral support, in combination with loads that are not in a plane of symmetry, such as is usually the case when the channel, angle, or other unsymmetrical section is used.† Special attention is given to the problem of combined bending and torsion, and a simplified procedure is presented for W beam shapes.

Composite design is mentioned briefly, but for detailed coverage the reader is referred to the AISCM, which provides a very adequate presentation of this topic.

10.2 TORSION

If torsion in a structural member is a major part of the load system, either a cylindrical or box tube should be used if possible. The cylindrical tube utilizes material in the most effective way possible for resistance to torsion.

---

†Refer to Section 3.1 for a review of support conditions required to permit use of simple bending theory in design.

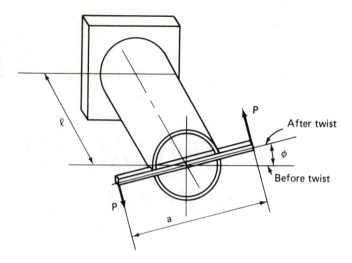

Fig. 10.1 Hollow cylinder in pure torsion.

Box shapes are a close second. The tube in Fig. 10.1 is loaded in pure torsion, as is the drive shaft of an automobile or the propeller shaft of a ship under its primary load. In Fig. 10.1, the torsional moment, M_t, is equal to Pa and the end twists through a total angle ϕ. For a tubular member under uniform pure torsion the angle of twist per unit length is constant:

$$\theta = \frac{\phi}{l} \tag{10.1}$$

The stress f_v in a torqued tube is "pure shear," as is indicated in Fig. 10.2.

Fig. 10.2 Section through a hollow cylinder in pure torsion.

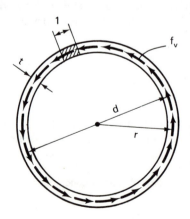

In a thin-walled tube the shear stress can be assumed constant through the wall thickness t, and each unit distance around the circumference exerts a tangential force equal to tf_v. The twisting moment about the central axis of the cylinder, at O, of each unit length of tangential shear force is $tf_v r$, where r is the mean radius of the cylinder. Summing up the contributions of each unit length of circumference, the total torsional moment is equal to $tf_v r$ multiplied by the circumferential length; hence

$$M_t = 2\pi t f_v r^2 \qquad (10.2)$$

The mean radius of the tube encompasses an area equal to

$$A_o = \pi r^2$$

Hence an alternative expression to Eq. (10.2) can be written

$$M_t = 2A_o t f_v \qquad (10.3)$$

The form of Eq. (10.3) is useful in that it applies to square and rectangular box tubes as well as cylindrical. If the wall thickness of a box section varies, Eq. (10.3) still applies, but t should be taken as the thickness of the thinnest plate segment, as this will be the most highly stressed and will determine the allowable torsional moment.

When a box section is twisted, plane sections remain plane—or very nearly so—after twist, and their contribution to the torsional resistance is in proportion to their distance from the center of twist. When an "open" section, such as a wide-flange shape, is twisted, elements not centered on the axis of twist *tilt* or *warp*, and unless such warping is in some way restrained the torsional resistance of each component is independent of its location in the cross section. Thus "closed" or box members are many times more rigid than "open" sections of the same general dimensions and weight per unit length. The torsional rigidity of a member is measured by the torsion constant J of the cross section, just as the bending rigidity is measured by the moment of inertia I. For a member under uniform torsion with no section restrained against warping, the general relationship between torsional moment and angle of twist per unit length is

$$M_t = JG\theta \qquad (10.4)$$

For a circular cross section, solid or hollow, J is equal to the polar moment of inertia; for any noncircular section it is always less than the polar moment of inertia. For a closed box section of *any* shape, enclosing only one internal cell,

$$J = \frac{4A_o^2}{\displaystyle\sum_{i=1}^{n} (s_i/t_i)} \qquad (10.5)$$

The denominator in Eq. (10.5) is the summation of the length/thickness ratios of all the n component parts of the tube around the periphery of the cross section. Thus for a thin-walled hollow cylinder, $A_o = \pi d^2/4$ and $\sum (s_i/t_i) = \pi d/t$, and for the circular pipe or tube

$$J = \frac{\pi d^3 t}{4} \tag{10.6}$$

For the box beam shown in Fig. 10.3, $\sum (s_i/t_i) = 2(h/t_w + b/t_f)$, and by Eq. (10.5),

$$J = \frac{2b^2 h^2}{(h/t_w) + (b/t_f)} \tag{10.7}$$

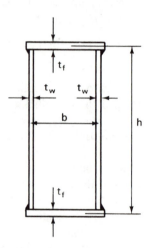

Fig. 10.3 Box-section nomenclature.

The torsion constant of a solid rectangular bar section, several times wider than its thickness, is approximately

$$J = \tfrac{1}{3} bt^3 \tag{10.8}$$

The torsion constant of open (i.e., nontubular) structural shapes, such as the wide-flange beam or angle, is simply approximated by summing Eq. (10.8) for the various component rectangular parts. More accurately, the AISCM lists J for standard shapes.

The maximum shear stress in a structural shape of open section under uniform torsion is

$$f_v = \frac{M_t t}{J} \tag{10.9}$$

In Eq. (10.9) the maximum shear stress is obviously located where the thickness t is greatest.

When flange warping is unrestrained, the distribution of shear stress in a wide-flange shape under uniform torsion is as shown in Fig. 10.4(a). When

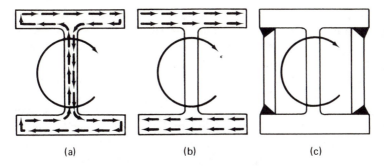

Fig. 10.4 Torsional shear stress distributions at unrestrained and restrained locations in a wide-flange beam.

warping is restrained, the shear stress is nearly constant through the flange thickness, as shown in Fig. 10.4(b). The flanges are then stressed in shear as they would be in a box girder.

Warping restraint may be obtained locally by adding longitudinal stiffeners as shown in Fig. 10.4(c). To be fully effective, these most be combined with lateral stiffeners at the ends of the longitudinal segments.

At sections away from a restrained location, the stress distribution is a combination of that shown in Fig. 10.4(a) and (b), approaching that of Fig. 10.4(a) as the distance from the restrained location increases. A *warping constant* C_w is also tabulated in the AISCM. Another torsion constant, the *torsion bending constant*, a, gives a rough approximation of the distance along a beam away from a restrained location that is required for the effect of restraint to be dissipated and the condition of Fig. 10.4(a) to be approached. The torsion bending constant may be calculated readily from the listed values of J and C_w in the AISCM.

$$a = 1.61\sqrt{\frac{C_w}{J}} \qquad (10.10)$$

The dimensionless parameter l/a will be used in the next section in a simplified presentation of combined bending and torsion with comparative design examples involving both box and open sections.

10.3 COMBINED BENDING AND TORSION

In the structural design of buildings it is desirable and usually possible to avoid the complications of torsion by proper location of members. But if torsion cannot be avoided in a laterally loaded beam, the problem involves *combined bending and torsion*, and the rectangular box section should be used if feasible. The AISCM provides dimensions and properties of available pipe and both square and rectangular box tubing.

In designing a box beam for combined bending and torsion, the preliminary selection may be made for bending moment alone, with a subsequent check on the combined shear stress due to both bending and torsion. More often than not the preliminary selection will be adequate. Where local concentration of torque load is introduced, care should be taken to guard against local distortion of the cross section. External stiffeners or internal diaphragms may be needed. If the framing is conducive to the use of a triangular closed box section, cross-sectional distortion is automatically eliminated.

For very short and stubby members, a W section may be suitable in combined bending and torsion. If the length l of a cantilever beam is less than $0.5a$ [Eq. 10.10)], the individual flanges may be assumed to take all the torsional moment as if they were individual cantilever beams, loaded in opposite directions. The cantilever beam shown in Fig. 10.5(a) is to be designed for a cantilever bending moment equal to Pl combined with a torsional moment equal to Pe. These may be considered separately for the loads shown in Fig.

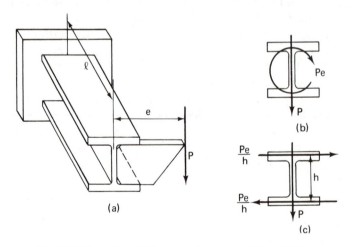

Fig. 10.5 Very short W beam section assumed to resist torsion by flange shear forces.

10.5(b) and, for the relatively short beam with l/a less than 0.5, as shown in Fig. 10.5(c). If the ratio $C_n = S_x/S_y$ can be estimated (see the discussion in Section 3.8 on biaxial bending), a preliminary estimate of the required section modulus can be made. The section modulus of an individual flange of a W section is approximately $S_y/2$. Thus for bending and torsional moments,

$$M_x = Pl \quad \text{and} \quad M_f = \frac{Pel}{h}$$

and the required section modulus is

$$S_x = \frac{M_x + 2C_n M_f}{F_b} \qquad (10.11)$$

For cantilever loads producing bending about the weak bending axis a more direct selection may be made, the required section modulus about yy being

$$S_y = \frac{M_y + 2M_f}{F_b} \qquad (10.12)$$

Simplified formulas for the stress in very long W beams can be written; but unless the maximum permissible angle of twist is very large, a tubular member will be desirable. Tables 10.1 and 10.2 tabulate simple approximate formulas for maximum flange moment and maximum angle of twist for the six cases of combined bending and torsion that are most commonly encountered. The formulas include simple approximations for short and long beams and interpolation formulas for intermediate-length beams. The principal advantage of these formulas over the exact ones is the fact that they may be used for direct design checks without reference to tables of hyperbolic or exponential functions. Formulas for the maximum shear stress due to torsion are not tabulated.

For short beams in the minimum l/a category, the shear stress will be mostly of the type shown in Fig. 10.4(b). Neglecting the shear stress of the type in Fig. 10.4(a), the maximum shear stress may be assumed to be 1.5 times the average at the location along the beam where shear due to torsional flange bending is greatest. To calculate the flange shear due to torsional flange bending, each individual flange may be assumed to act as a beam, loaded as shown in Fig. 10.5(c), with load magnitudes equal to the total beam load times e/h. The load distributions and beam end conditions for the individual flanges in torsional bending will be identical with those shown in Tables 10.1 and 10.2 for bending of the complete beam. Likewise, the maximum beam moments and flange bending moments occur at identical locations, and the direct stresses in the flanges due to the two causes are additive. Thus Eqs. (10.11) and (10.12), as applied to the short cantilever, may also be applied to any of the other five loading and support conditions.

The design of intermediate and long W beams in combined bending and torsion will be feasible only if the torsional moments are small. Otherwise, torsional deflections will be prohibitively large. Direct flange stresses should be checked for combined bending and torsion, and the maximum twist determined. Away from restrained ends the torsional shear stress, of the pattern shown in Fig. 10.4(a), may be calculated by Eq. (10.9), but it is bound to be small and of little design significance if the total twist angles are small.

Table 10.1

Approximate Formulas for Flange Moment and Total
Twist Angle: Three Concentrated Load Cases, 1, 2, and 3

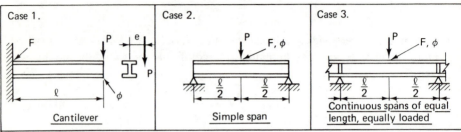

Case 1.	Case 2.	Case 3.
Cantilever	Simple span	Continuous spans of equal length, equally loaded

Maximum lateral bending moment in each flange at location F.

ℓ/a less than 0.5:	ℓ/a less than 1.0:	ℓ/a less than 2.0:
$M_f = \dfrac{Pe\ell}{h}$	$M_f = \dfrac{Pe\ell}{4h}$	$M_f = \dfrac{Pe\ell}{8h}$
ℓ/a more than 2.0:	ℓ/a more than 4.0:	ℓ/a more than 8.0
$M_f = \dfrac{Pea}{h}$	$M_f = \dfrac{Pea}{4h}$	$M_f = \dfrac{Pea}{8h}$

Maximum total angle of twist at end of a cantilever or a midspan of other beams, at location ϕ.

ℓ/a less than 0.5:	ℓ/a less than 1.0:	ℓ/a less than 2.0:
$\phi_t = 0.32\dfrac{Pea}{JG}\left(\dfrac{\ell}{a}\right)^3$	$\phi_t = 0.32\dfrac{Pea}{JG}\left(\dfrac{\ell}{2a}\right)^3$	$\phi_t = 0.64\dfrac{Pea}{JG}\left(\dfrac{\ell}{2a}\right)^3$
ℓ/a more than 2.0:	ℓ/a more than 4.0:	ℓ/a more than 8.0:
$\phi_t = \dfrac{Pe}{JG}(\ell - a)$	$\phi_t = \dfrac{Pe}{JG}\left(\dfrac{\ell}{2} - a\right)$	$\phi_t = \dfrac{2Pe}{JG}\left(\dfrac{\ell}{4} - a\right)$

Interpolation formulas for bending moment in each flange at F and maximum total twist angle at location ϕ.*

Case 1.   Cantilever. ℓ/a more than 0.5 and less than 2.0 :

$$M_f = \frac{Pea}{h}\left[0.05 + 0.94\left(\frac{\ell}{a}\right) - 0.24\left(\frac{\ell}{a}\right)^2\right], \phi_t = \frac{Pea}{JG}\left[-0.029 + 0.266\left(\frac{\ell}{a}\right)^2\right]$$

Case 2.   Simple Span. ℓ/a more than 1.0 and less than 4.0 :

$$M_f = \frac{Pea}{2h}\left[0.05 + 0.94\left(\frac{\ell}{2a}\right) - 0.24\left(\frac{\ell}{2a}\right)^2\right], \phi_t = \frac{Pea}{JG}\left[-0.029 + 0.266\left(\frac{\ell}{2a}\right)^2\right]$$

Case 3.   Continuous spans of equal length, equally loaded.
  ℓ/a more than 2.0 and less than 8.0 :

$$M_f = \frac{Pea}{2h}\left[0.05 + 0.94\left(\frac{\ell}{4a}\right) - 0.24\left(\frac{\ell}{4a}\right)^2\right], \phi_t = \frac{2Pea}{JG}\left[-0.029 + 0.266\left(\frac{\ell}{4a}\right)^2\right]$$

*Note.   At the extreme range of application the interpolation formulas are more accurate than the formulas given for locations outside of this range, yielding values slightly less than the others, which err slightly on the side of conservative design estimates.

Table 10.2

Approximate Formulas for Flange Moment and Total
Twist Angle: Three Uniform Load Cases, 4, 5, and 6

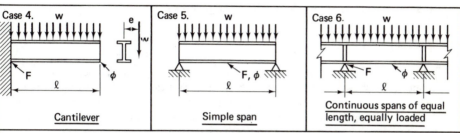

Case 4.	Case 5.	Case 6.
Cantilever	Simple span	Continuous spans of equal length, equally loaded

Maximum lateral bending moment in each flange at location F.

ℓ/a less than 0.5 :	ℓ/a less than 1.0 :	ℓ/a less than 2.0 :
$M_f = \dfrac{w\ell^2 e}{2h}$	$M_f = \dfrac{w\ell^2 e}{8h}$	$M_f = \dfrac{w\ell^2 e}{12h}$
ℓ/a more than 3.0 :	ℓ/a more than 6.0 :	ℓ/a more than 8.0 :
$M_f = \dfrac{w\ell ea}{h}\left(1 - \dfrac{a}{\ell}\right)$	$M_f = \dfrac{wea^2}{h}$	$M_f = \dfrac{w\ell ea}{h}\left(\dfrac{1}{2} - \dfrac{a}{\ell}\right)$

Maximum total angle of twist at end of a cantilever or at midspan of other beams, at locations ϕ.

ℓ/a less than 0.5 :	ℓ/a less than 1.0 :	ℓ/a less than 2.0 :
$\phi_t = 0.114 \dfrac{w\ell ea}{JG}\left(\dfrac{\ell}{a}\right)^3$	$\phi_t = 0.094 \dfrac{w\ell ea}{JG}\left(\dfrac{\ell}{2a}\right)^3$	$\phi_t = 0.151 \dfrac{w\ell ea}{JG}\left(\dfrac{\ell}{4a}\right)^3$
ℓ/a more than 3.0 :	ℓ/a more than 6.0 :	ℓ/a more than 8.0 :
$\phi_t = \dfrac{w\ell ea}{JG}\left(\dfrac{\ell}{2a} - 1 + \dfrac{a}{\ell}\right)$	$\phi_t = \dfrac{w\ell ea}{JG}\left(\dfrac{\ell}{8a} - \dfrac{a}{\ell}\right)$	$\phi_t = \dfrac{w\ell ea}{JG}\left(\dfrac{\ell}{8a} - \dfrac{1}{2}\right)$

Interpolation formulas for bending moment in each flange at F and maximum total twist angle at location ϕ.*

Case 4. Cantilever. ℓ/a more than 0.5 and less than 3.0 :

$$M_f = \frac{w\ell ea}{h}\left[0.041 + 0.423\frac{\ell}{a} - 0.068\left(\frac{\ell}{a}\right)^2\right]$$

$$\phi_t = \frac{w\ell ea}{JG}\left[-0.023 + 0.029\frac{\ell}{a} + 0.086\left(\frac{\ell}{a}\right)^2\right]$$

Case 5. Simple span. ℓ/a more than 1.0 and less than 6.0 :

$$M_f = \frac{w\ell ea}{h}\left[0.097 + 0.094\frac{\ell}{2a} - 0.0255\left(\frac{\ell}{2a}\right)^2\right]$$

$$\phi_t = \frac{w\ell ea}{JG}\left[-0.032 + 0.062\frac{\ell}{2a} + 0.052\left(\frac{\ell}{2a}\right)^2\right]$$

Case 6. Continuous spans of equal length, equally loaded.
ℓ/a more than 2.0 and less than 8.0 :

$$M_f = \frac{w\ell ea}{h}\left[0.005 + 0.342\frac{\ell}{4a} - 0.078\left(\frac{\ell}{4a}\right)^2\right]$$

$$\phi_t = \frac{w\ell ea}{2JG}\left[-0.029 + 0.266\left(\frac{\ell}{4a}\right)^2\right]$$

*Note. See footnote to Table 10.1

Several design examples will now illustrate the foregoing procedures and indicate situations where W shapes are satisfactory in combined bending torsion, as well as other cases where the reverse is true.

Example 10.1

A 3-kip pull is applied at any angle, tangential to the circumference of a 20-in. diameter, as shown, at the top of a 20-in.-long W beam that is fixed at its base. Select beam size using A36 steel, $F_y = 36$ ksi. Obviously, the design may be based on the pull being directed so as to cause bending about the weak axis of the beam, as shown at the bottom of the accompanying drawing.

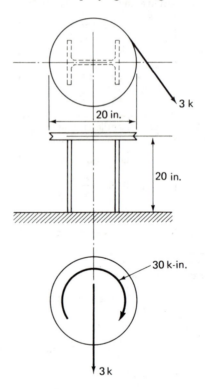

Solution

Refer to case 1, Table 10.1, and assume that l/a is less than 0.5—to be checked after selection. Try a W 12 beam and assume that $h = 11.5$ in.

$$M_f = \frac{Pel}{h} = \frac{3 \times 10 \times 20}{11.5} = 52.0 \text{ kip-in.}$$

$$M_y = Pl = 3 \times 20 = 60.0 \text{ kip-in.}$$

By Eq. (10.12), the required section modulus S_y is determined:

$$S_y = \frac{60.0 + 2 \times 52.0}{24} = 6.83 \text{ in.}^3$$

Try W 12 × 35:

$$S_y = 7.47 \text{ in.}^3$$
$$J = 0.74 \text{ in.}^4$$
$$C_w = 879 \text{ in.}^6$$
$$h = 12.50 - 0.52 = 11.98 \text{ in.}$$

Check l/a by Eq. (10.10):

$$b_f = 6.56 \text{ in.}$$

$$a = 1.61 \sqrt{\frac{879}{0.74}} = 55.5 \text{ in.}$$

$$\frac{l}{a} = \frac{20}{55.5} = 0.36 < 0.50 \quad \text{OK}$$

Check the direct stress in the flange:

$$f_b = \frac{60.0}{7.47} + \frac{2 \times 3 \times 10 \times 20}{11.98 \times 7.47} = 21.4 \text{ ksi} < 24 \quad \text{OK}$$

Check the maximum shear stress:

$$A_f = 6.56 \times 0.52 = 3.41 \text{ in.}^2$$

$$V_{\text{max}} = \frac{3}{2} + \frac{3 \times 10}{11.98} = 4.00 \text{ kips}$$

$$f_v = \frac{1.5 \times 4.00}{3.41} = 1.76 \text{ ksi} < 14.5 \quad \text{OK}$$

Note that in spite of the very short length of the beam and the orientation that makes the direct flange shear forces due to torsion and bending directly additive, the maximum shear stress is not significant.

Example 10.2

Redesign for Example 10.1, changing W section to a pipe section. Ignore torsion in preliminary selection and assume that $F_b = 22$ ksi.

Solution

The section modulus required is

$$S = \tfrac{60}{22} = 2.73 \text{ in.}^3$$

Try a 4-in. standard pipe (refer to AISCM):

$$S = 3.21 \text{ in.}^3 \qquad \text{O.D.} = 4.5 \text{ in.}$$
$$A = 3.17 \text{ in.}^2 \qquad \text{mean diameter: } d_m = 4.5 - 0.237 = 4.26 \text{ in.}$$
$$t = 0.237 \text{ in.}$$

Check the shear stress due to torsion and bending. (The shear shape factor for a circular tube in bending is 1.33.)

Due to bending:

$$f_v = \frac{1.33 \times 3}{3.17} = 1.26 \text{ ksi}$$

Due to torsion:

$$f_v = \frac{3 \times 10}{2 \times 0.237 \times \pi \times 2.13^2} = 4.44 \text{ ksi} \qquad \text{(Eq. 10.3)}$$

Due to both bending and torsion:

$$f_v = 1.26 + 4.44 = 5.7 \text{ ksi} < 14.5 \qquad \text{OK}$$

Direct stress due to bending:

$$f_b = \frac{60}{3.21} = 18.7 \text{ ksi}$$

Reduced allowable direct stress because of shear stress [Eq. (10.13)]†:

$$F_b = F_{rt} = \left[1 - \left(\frac{4.44}{22}\right)^2\right]22 = 21.1 \text{ ksi} > 18.7 \qquad \text{OK}$$

Examples 10.1 and 10.2 showed that either a W or pipe section could serve satisfactorily in combined bending and torsion for a short cantilever member. The required W section weighed three times more than the pipe. In the design of relatively long members in combined bending and torsion, such as might be required for a highway direction sign attached to a single vertical member, the pipe is the only suitable member because of the unduly large torsional deflections that would result from use of a W section.

Examples 10.3 and 10.4 illustrate the combined bending and torsion problem in the design of a continuous spandrel beam supporting an exterior wall. It is assumed that the framing situation does not permit reduction or elimination of the torsion component by means of a laterally contiguous floor slab or framed beams.

Example 10.3

Continuous 24-ft spans of a spandrel beam carry a wall load of 600 lb/ft, 4 in. from the center of the beam. Use A36 steel and a W beam. Wall support brackets will not be considered as adding to the beam section.

---

†*Note regarding combined shear and direct stress:* The AISCS does not provide any limitation on direct stress combined with shear, except in the case of connection design. To keep the maximum principal stress less than F_t, the allowable direct stress, the solution of the quadratic principal stress formula may be avoided for the simple case of a single direct stress component by the determination of a reduced allowable direct stress (F_{rt}) as follows:

$$F_{rt} = \left[1 - \left(\frac{f_v}{F_t}\right)^2\right]F_t \qquad (10.13)$$

If f_v/F_t is less than 0.2, the effect of shear stress on the allowable stress may be ignored. The case in Example 10.2 is borderline.

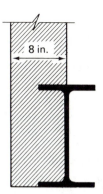

Solution

A preliminary trial selection will be chosen by designing for bending alone, but at a greatly reduced allowable stress, say, one-third of 22, or 7.33 ksi. Because of lack of lateral support, the provisions of AISCS, Sec. 1.5.1.4.1, for reduced end moment in the continuous beam cannot be applied. Design moment, assuming a beam weight of 30 lb/ft, is calculated:

$$M_x = \frac{0.63 \times 24^2 \times 12}{12} = 362.9 \text{ kip-in.}$$

Required section modulus for trial beam at $\frac{F_b}{3} = 7.33$ ksi,

$$S_x = \frac{362.9}{7.33} = 49.5 \text{ in.}^3$$

Refer to AISCM and choose a W 14 × 38 as a trial, for which

$$S_x = 54.6 \text{ in.}^3$$
$$S_y = 7.88 \text{ in.}^3$$
$$J = 0.80 \text{ in.}^4$$
$$C_w = 1230.0 \text{ in.}^6$$
$$h = 14.10 - 0.51 = 13.59 \text{ in.}$$

By Eq. (10.10),

$$a = 1.61\sqrt{\frac{1230}{0.80}} = 63.1 \text{ in.}$$

$$\frac{l}{a} = \frac{288}{63.1} = 4.56$$

(More than 2.0, less than 8.0; hence use interpolation formulas of Table 10.2, case 6.) Using the interpolation formula, the maximum moment in a flange due to torsional bending is found:

$$M_f = \frac{(0.64)(288)(4.0)(63.1)}{(12)(13.59)}\left[0.005 + 0.342\left(\frac{4.56}{4}\right) - 0.078\left(\frac{4.56}{4}\right)^2\right] = 83.7 \text{ kip-in.}$$

The stress due to torsional bending alone is

$$f_{bt} = \frac{83.7 \times 2}{7.88} = 21.2 \text{ ksi}$$

too great, obviously, for the torsion bending stress alone. To make a better selection, assume $M_f = 84$ kip-in., as in this case, $S_x/S_y = 54.7/7.88 = 6.93$. Then, by (Eq. 10.11),

$$\text{req'd } S_x = \frac{362.9 + 2 \times 6.96 \times 84}{22} = 69.6 \text{ in.}^3$$

requiring, by AISCM, a W 21 × 44. But, noting that S_x/S_y would be more than 10, the requirement for S_x would be escalated. It is obviously desirable to stay with a wider and less deep cross section. Noting, also, that the next group of sections heavier than the W 14 × 30 trial selection have $S_x/S_y = 5.5$ for the median of the group (AISCM), the required S_x would be reduced to

$$\frac{362.9 + 2 \times 5.5 \times 84}{22} = 58.5 \text{ in.}^3$$

Try a W 14 × 43, for which $S_x = 62.7$ in.^{3}
  Other needed properties of the W 14 × 43 are

$$S_y = 11.3 \text{ in.}^3$$

$$J = 1.05 \text{ in.}^4$$

$$C_w = 1950.0 \text{ in.}^6$$

$$h = 13.66 - 0.53 = 13.13 \text{ in.}$$

$$a = 1.61\sqrt{\frac{1950}{1.05}} = 69.4 \text{ in.}$$

$$\frac{l}{a} = \frac{288}{69.4} = 4.15$$

Again, use interpolation formulas from Table 10.2, case 6, for flange moment:

$$M_f = \frac{(0.64)(288)(4)(69.4)}{(12)(13.13)}\left[0.005 + 0.342\left(\frac{4.15}{4}\right) - 0.078\left(\frac{4.15}{4}\right)^2\right] = 89.5 \text{ kip-in.}$$

Stress due to bending moment M_x:

$$f_{bx} = \frac{0.64 \times 24^2 \times 12}{12 \times 62.7} = 5.9 \text{ ksi}$$

Stress due to torsional flange bending:

$$f_{bt} = \frac{89.5 \times 2}{11.3} \, 15.8 \text{ ksi}$$

Total direct stress due to combined bending and torsion:

$$f_b = 5.9 + 15.8 = 21.7 \text{ ksi} < 22 \quad \textbf{OK}$$

provided that 5.9 ksi is less than permissible stress for the laterally unsupported beam. Assume that $C_b = 1$ [AISCS, Sec. 1.5.1.4.5(2)]:

$$\frac{d}{A_f} = 3.22 \quad \text{(AISCM)}$$

F_b is no less than

$$\frac{12,000}{288 \times 3.22} = 12.94 > 5.9 \quad \textbf{OK}$$

  Check maximum twist at center by interpolation formula from Table 10.2, case 6:

$$\phi_t = \frac{(0.64)(288)(4)(69.4)}{(12)(2)(1.05)(11,200)}\left[-0.029 + 0.266\left(\frac{4.15}{4}\right)^2\right] = 0.047 \text{ radian}$$

Thus, if the masonry wall lacked internal restraint (if placed to a height of 50 in. before setting up of mortar), a point 50 in. above the top of the beam (assuming the beam to be about half loaded) would tend to deflect outward:

$$\tfrac{1}{2} \times 0.047 \times 50 = 1.18 \text{ in.}$$

Since the deflection would increase gradually as the wall was placed, it might be partially offset by progressive correction as bricks or blocks were placed. Thus, for eccentricities of *only an inch or two*, the use of W sections for spandrel beams, designed as above for combined bending and torsion, would be feasible without excessive twist.

Example 10.4

Alternative design using a box (rectangular tube) section for the same support and load conditions as Example 10.3. See the sketch.

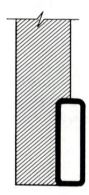

Solution

Initial design will be for full bending moment at $F_b = 22$ ksi, neglecting torsion. Assume that the weight of the member is 0.02 kip/ft

$$M_x = \frac{0.62 \times 24^2 \times 12}{12} = 375.1 \text{ kip-in.}$$

The required section modulus is

$$S_x = \frac{357.1}{22} = 16.23 \text{ in.}^3$$

Refer to the AISCM, for properties of rectangular structural tubing. Try TS 12 × 4 × 0.250:

$$S_x = 20.5 \text{ in.}^3 \qquad f_b = \frac{357.1}{20.5} = 17.4 \text{ ksi}$$

$$t = 0.25 \text{ in.}$$

$$h = 12.0 - 0.25 = 11.75 \text{ in.}$$

$$b = 4.0 - 0.25 = 3.75 \text{ in.}$$

$$A_o = 3.75 \times 11.75 = 44.1 \text{ in.}^2$$

Check the shear stress due to combined bending and torsion. Due to beam bending:

$$f_{vb} = \frac{0.61 \times 12}{0.5 \times 12} = 1.24 \text{ ksi}$$

Due to torsion:

$$M_t = \frac{wel}{2} = \frac{0.62 \times 4 \times 288}{12 \times 2} = 29.8 \text{ kip-in.}$$

$$f_{vt} = \frac{29.8}{2 \times 44.1 \times 0.25} = 1.35 \text{ ksi}$$

$$f_v = 1.24 + 1.35 = 2.6 \text{ ksi} < 14.5 \qquad \text{OK}$$

Check for lateral buckling. AISCS, Sec. 1.5.1.4.4, allows $F_b = 0.6F_y$ if depth is less than six times the width and $b/t < 238/\sqrt{F_y}$. (AISCS, Sec. 1.9.2.2 and Appendix A, Table 6)

$$\frac{d}{b} = 3 < 6 \qquad \text{and} \qquad \frac{b}{t} = 16 < 39.7 \qquad \text{OK}$$

Calculate the maximum twist angle at the center of span. The torsion constant, by Eq. (10.7), is

$$J = \frac{4 \times 44.1^2}{2(11.75 + 37.5)/0.25} = 62.7 \text{ in.}^4$$

Contrast this with $J = 1.05$ for the W 14 $\times$ 43 beam selection in Example 10.3. The *average* torsional moment between one end and the center of the span is $wel/4$ and

$$\phi_t = (\theta_{av})\left(\frac{l}{2}\right) = \frac{wel^2}{8JG}$$

$$= \frac{0.62 \times 4 \times 288^2}{12 \times 8 \times 62.7 \times 11{,}200} = 0.003 \text{ radian}$$

to be contrasted with ϕ_t of 0.047 radian for the W beam selection of Example 10.3.

The foregoing examples have demonstrated design procedures for both open and closed sections and have shown the strength and stiffness advantages of the closed box section in resisting torsion loads.

10.4   BIAXIAL BENDING
AND LATERAL-TORSIONAL BUCKLING

Properties for structural sections as listed in the AISCM are in terms of the *xx* and *yy* axes and, except for the angle and zee† sections, the *xx* and *yy* axes are also the *principal* axes of the cross sections, as will always be the case if one of the two axes is an axis of symmetry. As explained in Chapter 3, if

†No longer listed in the AISCM.

lateral support is provided, either continuously or at the locations of applied concentrated load, a beam of any shape may be designed on the basis of simple bending theory. If lateral support is not provided, the possibility of lateral-torsional buckling about the weakest principal axis is always present, and the design procedure to be recommended here requires the determination of the orientation of the principal axes, in cases where they are not already known, and the calculation of the principal moments of inertia, designated herein as I_1 and I_2. After these are determined, and the effect of lateral-torsional buckling in reducing the allowable stress in bending about the strong bending axis is estimated, the procedure parallels that of Chapter 3.

In the general case, to determine the principal moments of inertia, it will be necessary to calculate the product of inertia, I_{xy}. The reader should refresh his acquaintance with this parameter by reference to his text on strength of materials. It will be recalled that

$$I_x = \int y^2 \, dA \qquad I_y = \int x^2 \, dA \qquad I_{xy} = \int xy \, dA$$

Note that I_x and I_y are always positive quantities, but that I_{xy} may be either positive or negative, and that for the same section this will depend on the arbitary way in which the positive directions of x and y are chosen. If, as is so common, the structural shape is made of rectangular component parts, the contribution of any one rectangle to I_{xy} may be determined by the *parallel-axis theorem*,

$$I_{xy} = I_{xyo} + Ax_o y_o \qquad (10.14)$$

If the x and y axes include one that is an axis of symmetry, they are then principal axes, and I_{xyo} is zero. Thus, for a rectangular component, I_{xyo} is different from zero only if the sides are tilted at an angle to the x and y axes, as illustrated in Fig. 10.6, which is explanatory of Eq. (10.15):

$$I_{xyo} = \frac{b^3 t - b t^3}{12} \sin \theta \cos \theta \qquad (10.15)$$

If b is relatively large in relation to t, the term bt^3 may be omitted from Eq. (10.15) with but little error. It should be noted that when θ is either zero or

Fig. 10.6

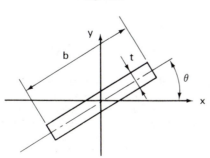

$90°$, I_{xyo} is zero. Another useful relationship is the fact that, regardless of the orientation of x and y, the sum of I_x and I_y is a constant. Thus, also,

$$I_x + I_y = I_1 + I_2 \tag{10.16}$$

Equation (10.16) is useful in the determination of the principal moments of inertia of an angle, using the information listed in the AISCM, as illustrated by Example 10.5.

Example 10.5

Determine the principal moments of inertia of an L $6 \times 4 \times \frac{1}{2}$.

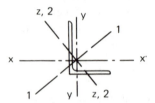

Solution

Referring to the AISCM,

$$I_x = 17.4 \text{ in.}^4 \qquad I_y = 6.27 \text{ in.}^4$$

and the minimum radius of gyration about the principal axis zz is listed as 0.870 in. The area A is given as 4.75 in.^{2}, and from the relationship $I = Ar^2$, the minimum moment of inertia, I_2, is

$$I_2 = 4.75 \times 0.87^2 = 3.60 \text{ in.}^4$$

Equation (10.16) now provides the calculation of I_1:

$$I_1 = 17.4 + 6.27 - 3.6 = 20.1 \text{ in.}^4$$

The complete problem of calculating section properties in the more general case for which no handbook information is available will be illustrated in Example 10.6. If the orientation of the principal axes is not known, it may be determined by Eq. (10.17), in which θ is the angle between the x axis and the principal axes. Only one of the angles need be calculated since they are $90°$ apart:

$$\tan 2\theta = \frac{2I_{xy}}{I_y - I_x} \tag{10.17}$$

The magnitudes of the principal moments of inertia are given by

$$I_1, I_2 = \frac{I_x + I_y}{2} \pm \sqrt{\left(\frac{I_x - I_y}{2}\right)^2 + I_{xy}^2} \tag{10.18}$$

Stresses may now be determined by resolving the loads into components in the principal planes and superposing the stresses as calculated by the ordinary beam stress formula applied successively to the bending moments about each of the two principal axes, as in the case treated in Chapter 3 [Eq. (3.12)], which was applicable when the handbook x and y axes were also principal axes. A scale layout will be helpful to determine distances to extreme fibers in the stress calculations.

However, if there is no lateral support, as is always the case if applied loads cause stress about both principal axes, there will be a reduction in allowable stress required for bending about the strong axis because of the lateral-torsional buckling effect. The AISCS interaction procedure, also explained and illustrated in Chapter 3 [Eq. (3.13)], may then be applied, provided that the reduced allowable stress for bending about the strong axis is determined, as will be discussed later.

Example 10.6

Determine the section properties about the x and y axes, and about the principal axes, for the shape shown. As a close approximation in the calculations it is assumed that the section is made up of two rectangular parts with breadths of 8 and 10 in., respectively.

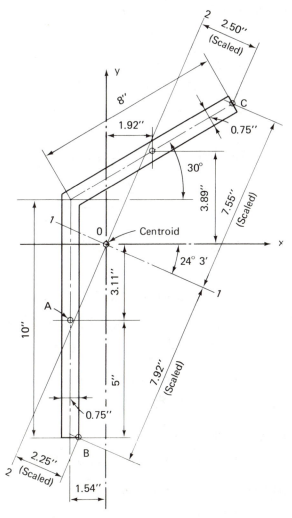

Solution

The area is

$$A = 0.75(8.0 + 10.0) = 13.5 \text{ in.}^2$$

Locate the neutral axes. Use point A, the centroid of the 10-in. segment, as a reference origin:

$$\bar{x}_A = \frac{6.0 \times 3.46}{13.5} = 1.54 \text{ in.}$$

$$\bar{y}_A = \frac{6.0 \times 7.0}{13.5} = 3.11 \text{ in.}$$

Having located the centroid, this becomes the origin of x and y distances to be used in calculation of the moments of inertia, I_x, I_y, and I_{xy}.

Determine I_x:

$$6.0 \times 3.89^2 + \frac{6.0 \times 4.0^2}{12} = 98.79$$

$$7.5 \times 3.11^2 + \frac{7.5 \times 10.0^2}{12} = 135.06$$

$$I_x = \overline{233.85} \text{ in.}^4$$

Determine I_y:

$$6.0 \times 1.92^2 + \frac{6.0 \times 6.92^2}{12} = 46.06$$

$$7.5 \times 1.54^2 + \frac{7.5 \times 0.75^2\dagger}{12} = 18.14$$

$$I_y = \overline{64.20} \text{ in.}^4$$

Determine I_{xy}: Calculate I_{xyo} of the 8×0.75 in. segment, by Eq. (10.15), for $\theta = 30°$,

$$\sin \theta = 0.500$$

$$\cos \theta = 0.866$$

Equation (10.15), for calculation purposes, may be more conveniently expressed as

$$I_{xyo} = \frac{bt(b^2 - t^2)}{12} \sin \theta \cos \theta$$

The t^2 term will be included for sake of completeness:

$$I_{xyo} = \frac{6.0(8.0^2 - 0.75^2)}{12}(0.5 \times 0.866) = 13.73 \text{ in.}^4$$

Thus for the complete section, by Eq. (10.14),

$$I_{xy} = (7.5)(-3.11)(-1.54) + (6.0)(+3.89)(+1.92) + 13.73 = +94.47 \text{ in.}^4$$

$\dagger$This term could have been omitted with but little error.

Orientation of the principal axes is obtained by use of Eq. (10.17),

$$\tan 2\theta = \frac{2(+94.47)}{64.20 - 233.85} = -1.114$$

Therefore, $2\theta = -48°5'$; hence $\theta = -24°3'$ and

$$\sin \theta = -0.4075$$
$$\cos \theta = 0.9132$$

The principal moments of inertia are now calculated by Eq. (10.18):

$$I_1, I_2 = \frac{233.85 + 64.20}{2} \pm \sqrt{\left(\frac{233.85 - 64.20}{2}\right)^2 + 94.47^2}$$

$$I_1 = 275.99 \text{ in.}^4$$
$$I_2 = 22.07 \text{ in.}^4$$

Suppose, now, that lateral support is provided for the section of Example 10.6, either continuously, if the load is continuous, or by rods attached at all load locations, as shown in Fig. 10.7. With such support, bending is forced to be about the

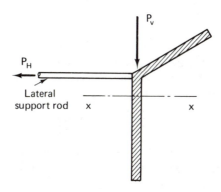

Fig. 10.7

xx axis and the stress may be calculated by the usual Mc/I formula. The load-carrying capacity is increased by such lateral support, and, if the use of supports is optional, the cost of supports can be weighed against the cost of a heavier member that would be required if such supports were lacking. The force P_H in the support may be calculated by

$$P_H = \frac{P_V I_{xy}}{I_x} \tag{10.19}$$

The design of a laterally supported unsymmetrical section is illustrated next.

Example 10.7

A member having the properties and cross section of Example 10.6 has a span of 18 ft and is loaded vertically at the third points. If the allowable stress is 22 ksi (A36 steel), what load can the member safely carry if lateral supports are provided at the load locations as shown in the sketch?

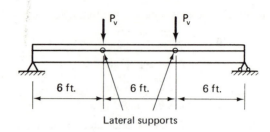

Lateral supports

Solution

Calculate the bending moment:

$$M_x = 72 \, P_V \text{ kip-in.}$$

Referring to Example 10.6, the maximum stress in tension, 8.11 in. below the neutral axis, will determine the load capacity:

$$f_b = 22 = \frac{72 \, P_V \times 8.11}{233.85}$$

Solving,

$$P_V = 8.81 \text{ kips}$$

In trying to bend laterally, about its weak axis, the member would obviously tend to deflect to the right. Thus the stress in the lateral support rods would be tension in the amount

$$P_H = \frac{8.81 \times 94.47}{233.85} = 3.56 \text{ kips}$$

If there is no lateral support, the allowable stress in bending about the strong axis will probably need to be reduced to provide safety against lateral-torsional buckling. The AISCS covers only specific cross sections. In the present case a conservative estimate of the critical moment that will cause lateral-torsional buckling is provided by

$$M_{cr} = \frac{\pi}{l} \sqrt{JGEI_2} \tag{10.20}$$

For steel, $\sqrt{GE} = 18{,}000$, and

$$M_{cr} = \frac{18{,}000\pi}{l} \sqrt{JI_2}$$

Although overconservative for bent beams, it is convenient in situations not covered by the specifications to convert the beam-buckling problem into an equivalent column problem and thus permit direct use of tables of column allowable stresses.

The buckling stress in compression is calculated:

$$f_{cr} = \frac{18{,}000\pi c_c}{lI_1} \sqrt{JI_2} \tag{10.21}$$

where c_c is the distance from the 1–1 principal axis to the extreme fiber in compression.

Table 9, Appendix A, of the AISCS, gives column buckling stresses, assuming elastic behavior, but divided by the long column factor of safety of $\frac{23}{12}$. Thus, if the beam buckling stress by Eq. (10.21) is divided by $\frac{23}{12}$, one can enter Table 9 with the corresponding stress and read out the "equivalent" column slenderness ratio, Kl/r. This can be used to determine a safe beam buckling stress by use of the allowable stress tables for columns, as given in the AISCM. The result is overconservative for two reasons: (1) Eqs. (10.20) and (10.21) neglect the bending contribution to torsional resistance, and (2) the factor of safety used in the column tables is greater than that specified for beams. The equivalent column procedure is now illustrated in Example 10.8.

Example 10.8

Same as Example 10.7, but without any lateral support.

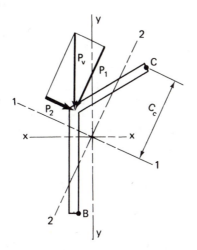

Solution

Resolve P_V into principal-plane components P_1 and P_2:

$$P_1 = P_V \cos \theta = 0.9132 P_V$$
$$P_2 = P_V \sin \theta = 0.4075 P_V$$

Stress at $B = 22$ ksi (trial):

$$22 = \frac{0.9132 P_V \times 6 \times 12 \times 7.92}{275.99} + \frac{0.4075 P_V \times 6 \times 12 \times 2.25}{22.07}$$

$$22 = 1.887 P_V + 2.991 P_V = 4.878 P_V$$

$$P_V = 4.51 \text{ kips}$$

Check the selection by the AISCS interaction formula procedure (refer to Chapter 3, Sec. 3.8). Referring to Example 10.6, the scaled distance from the 1–1 principal axis to the location of average maximum compression stress (C), distance c_c, is 7.55 in. The torsion constant, J, by Eq. (10.8), is

$$J = \frac{(10 + 8)\ 0.75^3}{3} = 2.53 \text{ in.}^4$$

By Eq. (10.21),

$$f_{cr} = \frac{18{,}000 \times 3.1416 \times 7.55}{216 \times 275.99}\sqrt{2.53 \times 22.07} = 53.5 \text{ ksi}$$

The equivalent column buckling stress, divided by $\frac{23}{12}$, is

$$F'_{e\,(equiv)} = \frac{53.5 \times 12}{23} = 27.91 \text{ ksi}$$

Referring to AISCS, Table 9, Appendix A, the equivalent

$$\frac{Kl}{r} = 73.1$$

Note: For remarks on load and support locations in Examples 10.7 and 10.8, refer to Section 10.5.

Now enter the Allowable Column Stress, Table 3-36, Appendix A, AISCS, and obtain

$$F_a = 16.10 \text{ ksi}$$

for a column with $Kl/r = 73.1$. This will be a safe allowable maximum stress for the beam buckling condition of this problem. It is obvious that the value of $P_V = 4.51$, based on a maximum stress of 22 ksi for bending about both axes, is too great. Try $P_V = 4$ kips and calculate bending stresses f_{b1} and f_{b2} separately:

$$f_{b1} = \frac{0.9132 \times 4 \times 72 \times 7.55}{275.99} = 7.19 \text{ ksi}$$

$$f_{b2} = \frac{0.4075 \times 4 \times 72 \times 2.50}{22.07} = 13.29 \text{ ksi}$$

Applying the interaction formula,

$$\frac{7.19}{16.10} + \frac{13.29}{22} = 1.05 > 1 \qquad \text{NG}$$

To avoid repeating the foregoing calculations, the load that will bring the right side of the interaction formula down to 1 can be approximated as equal to $4/1.05 = 3.81$ kips.

The approach used in Example 10.8 may also be applied to the design of box girders of such slender proportions as to fall outside the scope of the simplified rules suggested in Chapter 3, or to solid rectangular bars, or to other closed sections that fall outside specification rules. Example 10.9 will illustrate.

Example 10.9

Determine the allowable bending stress for a TS 12 × 2 × 0.250 in. rectangular tube used as a laterally unsupported simple beam with a span of 20 ft (A36 steel).

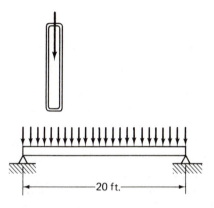

Solution

The AISCS, Sec. 1.5.1.4.4, permits a bending stress of $0.6F_y$ if "the depth is less than six times its width." For this section, $d/b = 6$, exactly borderline, but the example will serve to illustrate the procedure and give some indication as to the degree of conservatism that is involved in the specification.

The properties of this box tube are (AISCM)

$$I_x = 88.3 \text{ in.}^4$$
$$I_y = 4.51 \text{ in.}^4$$
$$S_x = 14.7 \text{ in.}^3$$

The torsion constant is calculated by Eq. (10.7):

$$J = \frac{2 \times 1.75^2 \times 11.75^2}{(11.75 + 1.75)/0.25} = 15.66 \text{ in.}^4$$

Calculate the critical stress by Eq. (10.21):

$$f_{cr} = \frac{18,000\pi}{240 \times 14.7}\sqrt{15.66 \times 4.51} = 134.7 \text{ ksi}$$

The equivalent column buckling stress, divided by $\frac{23}{12}$, is

$$F'_e = \frac{134.7 \times 12}{23} = 70.3 \text{ ksi}$$

and from AISCS, Appendix A, Table 9, the equivalent

$$\frac{Kl}{r} = 46.1$$

AISCS, Appendix A, Table 3-36, gives an allowable column stress $F_a = 18.69$ ksi. This is a very conservative estimate for the allowable maximum stress in bending in this example, hence cannot be considered to be in great disagreement with the fact that AISCS would allow 22 ksi in this borderline situation.

10.5 SHEAR CENTER

If certain structural sections, such as the channel and angle, are loaded through their centroidal axis without any torsional support or torsional restraint at the load points, they will twist. The design problem is then one of combined bending and torsion, as covered in Section 10.3, a complication that may be avoided if the member can be loaded and supported through its *shear-center axis*. In the case of the channel the three-dimensional free-body equilibrium diagram sketched in Fig. 10.8(a) illustrates a short segment, Δx in length, cut from a beam shown loaded through the shear-center axis, so as to avoid twist. Since the loads are parallel to a principal axis of the cross

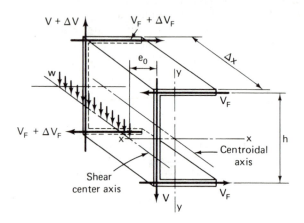

Fig. 10.8(a)

section, the member will be in simple bending and without twist. The torsional couple $\Delta V_F h$ is held in equilibrium by the opposed torsional couple $w(\Delta x)e_o$, as illustrated. Referring to Fig. 10.8(b), the distance from the middle plane of the web to the shear-center axis is

$$e_o = \frac{x_o h^2}{4r_x^2} \tag{10.22}$$

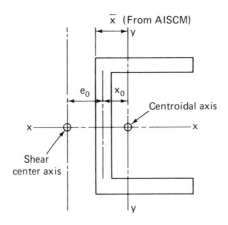

Fig. **10.8(b)** Shear center of channel.

where r_x is the radius of gyration about the xx axis. Equation (10.22) applies to channels with nonparallel flange faces as well as parallel.

If a channel supports beams that frame into it, the arrangement in Fig. 10.9(a) is preferred over that of Fig. 10.9(b). The channel may be assumed to be without twist and loaded through its shear-center axis, in either case, provided that the supported beams are designed for span L measured in each case from the shear-center axis. In addition, in Fig. 10.9(a) connecting bolts should be located as near as possible to a vertical line passing through the shear-center axis, whereas in Fig. 10.9(b) the connection should preferably be designed for an eccentricity of load equal to the distance from the shear-center axis to the bolt line. The channel as a spandrel beam loaded through the shear center is treated in Example 10.10.

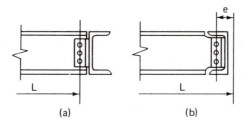

Fig. **10.9** Alternative framing arrangements for channel beam.

Example 10.10

Select a channel as a 22-ft simple span spandrel beam for a load of 2 kips/ft, including weight of channel, and locate 8-in. wall so as to eliminate torsion of beam.

Solution

The maximum bending moment is

$$M_x = \frac{2 \times 22^2 \times 12}{8} = 1425 \text{ kip-in.}$$

The required section modulus, assuming that $F_b = 22$ ksi, is

$$S_x = \frac{1425}{22} = 66 \text{ in.}^3$$

Referring to the beam selection tables, AISCM, MC 18 $\times$ 51.9 with $S_x = 69.7$ is OK, as is the lighter weight W 21 $\times$ 44, but if it is desired to hide the beam by the concrete wall, the channel offers advantages in the elimination of the combined bending and torsion problem, taking advantage of the shear-center location, as shown in the sketch. If laterally unsupported, d/A_f (AISCM) is 7.02, and the allowable stress is

$$F_b = \frac{12,000}{264 \times 7.02} = 6.48 \text{ ksi}$$

Thus, during construction, at least temporary lateral support would be needed. (Note that the beam selection tables list the maximum span, L_u, for which no lateral support is needed, in this case, 6.6 ft.) From AISCM, the properties of the MC 18 $\times$ 51.9 are

$$S_x = 69.7 \text{ in.}^3 \qquad r_x = 6.41 \text{ in.} \qquad \bar{x} = 0.858$$

In Eq. (10.22), $x_o = \bar{x} - (t_w/2) = 0.858 - 0.300 = 0.558$ in.

$$h = d - t_f = 18.0 - 0.625 = 17.375 \text{ in.}$$

By Eq. (10.22),

$$e_o = \frac{0.558 \times 17.375^2}{4 \times 6.41^2} = 1.02 \text{ in.}$$

suggesting the relative channel and wall location as shown in the sketch.

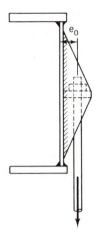

Fig. 10.10 Channel girder supporting hanger loads.

Another situation conducive to use of a channel beam may occur if loads must be suspended by hanger rods, as shown in Fig. 10.10. These may be located at the shear center and again the combined bending and torsion problem is avoided.

Formulas for the shear-center location of other shapes and procedures for determining the location for shapes having no axis of symmetry are given in Ref. 1.7. When a section consists of only two rectangular component parts, the shear center is at the intersection of their two middle planes, as shown for the angle and tee section in Fig. 10.11. Referring back to Examples 10.7 and 10.8, it should be noted that loads and supports were both introduced at the shear-center axis, thus eliminating torsion from the problem.

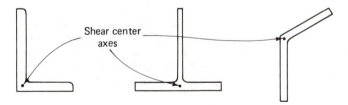

Fig. 10.11 Shear-center locations for two-part sections.

10.6 COMPOSITE DESIGN

In composite design two different materials are joined together so as to utilize the properties of each to the best advantage. Thus when a reinforced concrete floor slab is attached to a supporting system of steel beams by means of shear

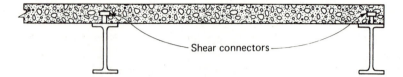

Fig. 10.12 Composite concrete slab and steel beam construction.

connectors, as shown in Fig. 10.12, the entire system acts as a unit. The slab
not only serves its usual purpose of supporting floor loads and transmitting
these to the beams, but also provides a compression flange augmenting the
top flange of the steel beams. The portion of the flange supplied by the concrete is either all or mostly in compression, utilizing concrete to its best
advantage.

The AISCM provides general notes, design examples, and composite
beam property tables for a wide variety of steel beams, both with and without
cover plates, and with slab thicknesses varying from 4 to $5\frac{1}{2}$ in. Quoting from
the *Manual*: "Composite construction is appropriate for any loading. It is
most efficient with heavy loading, relatively long spans, and beams spaced as
far apart as permissible." Composite design is widely used and offers advantages of economy and of overall structural integrity. Because of the excellent
coverage of the subject in the AISCM, further treatment in detail will herein
be omitted.

PROBLEMS†

10.1. Compare (a) torsional stiffness as measured by the torsion constant J for
the following sections, which have approximately the same cross-sectional
area, and (b) the torsional moment, for which the maximum shear stress is
equal to 15 ksi. In the case of the W section, calculate the torsion constant
by Eq. (10.8) and compare with the handbook value.
   a. Pipe 10 std. ($A = 11.9$ in.^{2})
   b. TS 10 × 10 × 0.3125 ($A = 11.7$ in.^{2})
   c. W 10 × 39 ($A = 11.5$ in.^{2})

10.2. A vertical pipe supports a sign, as shown, which is to be designed for a wind
force of 30 lb/ft^2. Neglecting the weight of the pipe and the sign, and neglecting the wind force on the pipe, select a size for the bending moment and
check the maximum shear stress due to combined torsion and bending.
Redesign if necessary.

†A36 steel unless otherwise specified.

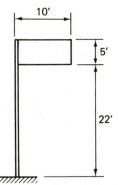

10.3. For the box section as shown in cross section, what is the torsional moment capacity for a maximum shear stress of 15 ksi? At this stress, what would the total angle of twist for a member 28 ft long amount to?

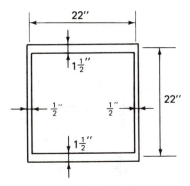

10.4. Using procedures applicable only to very short W sections, as described in connection with Fig. 10.5, determine the maximum normal and shear stress in the flange of the beam made up of three plates as shown.

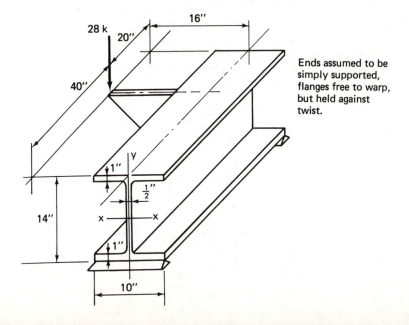

10.5. Similar to Example 10.3. Move the W section 2 in. laterally to reduce the eccentricity of load from 4 to 2 in. Design as a simple beam instead of continuous and reduce the span from 24 to 18 ft.

10.6. Redesign the situation in Problem 10.5 using a standard rectangular structural tube. Procedure is similar to that of Example 10.4.

10.7. Determine the principal moments of inertia of an L $8 \times 4 \times \frac{3}{4}$ by the procedure used in Example 10.5.

10.8. A zee bar, cross section as shown, is used as a simple beam, 18 ft between supports, and is loaded at the third points with loads of 1400 lb each. To prevent lateral movement and force bending to remain in the plane of the web, lateral ties are introduced at the third points as shown. Determine the maximum stress due to bending and the force in the lateral ties. In which direction is the tie force applied?

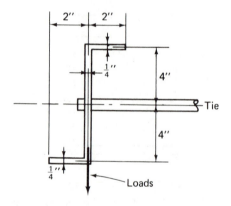

10.9. If the lateral ties are removed (refer to Problem 10.8), what is the maximum stress in the zee bar for the same loads that were applied in Problem 10.8?

10.10. Using a steel with $F_y = 50$ ksi, check the adequacy of the laterally unsupported zee section (refer to Problem 10.9) by use of the adaptation of the AISC interaction formula used in Example 10.8, utilizing the equivalent column allowable stress procedure for bending about the minor principal axis.

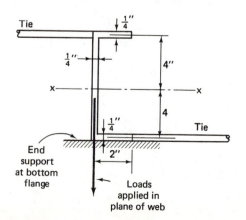

10.11. For a channel section with the same I_x and same area as the zee section of Problem 10.8, and with the same loads and span, ties must now be provided to prevent twisting, as shown, at the third points where the vertical loads are applied. For what force must the ties be designed and in what direction are the tie forces applied?

10.12. Design a welded channel girder of the type shown in Fig. 10.10 for a span of 36 ft to support third-point 60-kip loads $2\frac{1}{2}$ in. from the center line of the web with the channel proportioned so that the loads pass through the shear center to eliminate torsion. E70 electrodes and A36 steel may be assumed.

11

COMPUTER-AIDED TECHNOLOGY

11.1 INTRODUCTION

Since the early 1950s, use of the electronic digital computer has facilitated dramatic changes in all branches of engineering. Problems that were previously impossible or impractical by manual procedures can now be handled easily by use of the computer. The computer can be utilized to perform computations of great complexity in making design modifications, optimization, code checking, and so on, in a rapid, effective, and highly accurate manner.

Computer-aided technology (CAT) in design has achieved an engineer–computer interactive process which is now widely used in practice. The computer does all mathematical computations, conditional iterations, and code checking, while the design engineer exercises complete control over a variety of design decisions based on engineering judgment and experience, together with design constraints. Thus the system of man plus the computer is employed to achieve an optimum design solution.

The computer, of course, has been used since its initial availability in both university and industrial research. Emphasis herein is on design applications, an area of computer technology that has made especially rapid progress during the past several years.

11.2 BASIC FLOWCHART PROGRAMMING

The flowchart approach, introduced in earlier chapters as a guide to the use of the *AISC Design Specifications*, has had its primary application and origin as a basic aid in computer programming. It is a series of logical steps which direct the programmer along the flow path until the program objective is achieved. The integrated flowchart is primarily composed of the following three basic flow elements (modules): (1) sequential steps; (2) decision tests, providing direction to the next step to be executed as the result of a test; and (3) conditional iteration in meeting certain requirements and/or optimization. These three basic flow elements are integrated into a logical flow system which is not only used for computer programming, but is also extremely useful in the design office for guidance through the morass of the complex design codes and procedures. The entire flow process, then, becomes more simple, systematic, and manageable by means of the flowchart made up of the basic elements.

1. Sequential Steps

Logically independent sequential steps as shown in Flowchart 11.1 indicate that the control of the design process flows from box *A* to box *B*, and then on to *C*, and so on. The process boxes *A*, *B*, *C*, and so on, are all connected by one way direction flow arrows, and are explicitly representative of a set of requirements, assignments, or formulas that are to be executed by the computer subroutine (program module) and/or engineer.

Flowchart 11.1

Sequential Steps

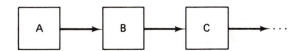

2. Decision Test

The *decision test symbol,* or *decision node,* indicates a test requirement as shown in Flowchart 11.2. The diamond-shaped test symbol is characterized by one flow-in arrow into the decision node and by two outflow arrows lead-

Flowchart 11.2

Decision Test Symbol

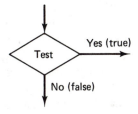

ing to the next appropriate step depending upon the result of a test requirement. The test requirement is a defined criterion to be met. Whether or not
the criterion is met determines which of the two alternative paths must be
followed in making an exit from that decision node.

3. Conditional Iteration

Conditional iteration is a higher-level module which is built up from the
first two types of modules. Conditional iteration provides a series of repetitive
executions (loop operations) dictated by the constraint parameters required
by codes and optimization. In the conditional iteration shown in Flowchart
11.3, the flow passes through the process box to the decision node, where a
test is made. If the test is no good (NG), the correction/optimization will be
made in favor of the test before returning back to the previous process. The
process is repeated until the test is met so that the flow can exit to a new
process.

Flowchart 11.3

Conditional Iteration

11.3   COMPUTER-AIDED DESIGN

Design is a process involving configuration, loadings, boundary conditions,
material properties, and specifications. A final design always results from a
sequence of problem-solving operations coupled with various optimizations.
In most situations, the design problem in steel does not have any easy or
direct closed-form solution at the beginning of the design. However, there are
some initial and preliminary solutions to initiate any complex design problem.
One of the most important of these design solutions is the use of CAT to
achieve a satisfactory solution in meeting the design criteria and performance
requirements. With the aid of a high-speed, electronic digital computer, the
integrated design system can be executed in part or in total with great effectiveness. The design system can be used as an analytical tool to allow rapid

synthesis by a number of successive iterations. The integrated system of computer-aided design is demonstrated in Flowchart 11.4, which consists of the following logical steps.

Flowchart 11.4

Computer-Aided Design System

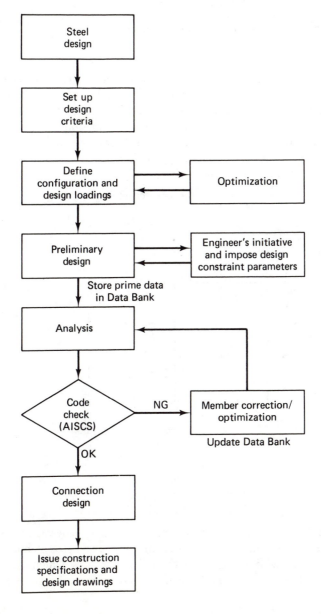

1. Design Criteria

One must establish complete design criteria before actual design work takes place. The design criteria provide a general design guide concerning the type of structural system, material strengths and grades, structure configuration, design loads, and specifications. The design-constraint parameters should be implemented together with the code-check requirements, which will be introduced subsequently.

2. Define Configuration and Design Loadings

Define Configuration

Structural configuration can be defined, in general, by three direction parameters, such as two dimensions in the north-south and east-west directions in the horizontal plan, and one dimension in the vertical direction (elevation). It is usually defined by column and floor spacings. Then, all joint coordinates can be generated automatically by computer whenever the spacings are given. Simultaneously, all joint identification names can be logically defined and composed of:

$$IJN \qquad \text{as illustrated in Fig. 11.1}$$

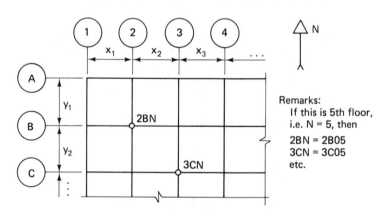

Fig. 11.1 Typical nth floor plan.

where

$$I = i\text{th column line in north-south direction}$$
$$J = j\text{th column line in east-west direction}$$
$$N = n\text{th floor of the joint}$$

Three-dimensional geometric optimization, in general, is required to determine an optimum number of floors for a given size of building. Configuration optimization minimizes the cost function ϕ_n, which consists of the primary cost parameters such as material, land, labor, design, maintenance, and so on, as follows:

$$\phi_n = f(x_1, x_2, x_3, x_4, x_5, \ldots, x_m)$$

where

$n =$ number of stories
$m =$ number of independent primary cost parameters
$x_1, x_2, x_3, x_4, x_5, \ldots$
$=$ independent primary cost parameters, such as material, land, labor, design, maintenance, etc.

To state the problem mathematically, then, the optimum conditions are defined as follows:

$$\frac{\partial \phi_n}{\partial x_1} = 0 \qquad \frac{\partial \phi_n}{\partial x_2} = 0$$

$$\frac{\partial \phi_n}{\partial x_3} = 0 \qquad \frac{\partial \phi_n}{\partial x_4} = 0$$

$$\frac{\partial \phi_n}{\partial x_5} = 0, \ldots, \frac{\partial \phi_n}{\partial x_m} = 0$$

In the computer-aided optimization, which will be introduced separately in Section 11.4, a simplified analytic model for a rapid configuration optimization can be developed in terms of the primary cost parameters mentioned for direct evaluation of a series of ϕ_n data, as illustrated in Fig. 11.2.

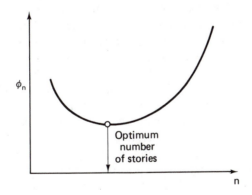

Fig. 11.2 Simplified analytical model for configuration optimization.

Design Loadings

There are several types of primary loadings that should be considered in design; they are dead, live, wind, and seismic loads. Dead loads can be either defined manually or obtained by computer when member sizes are known. Live loads include impact, snow, thermal, and any loads that are not permanently applied to the structure. Live loads shall be considered as applied either to the entire supporting area or to a portion of the supporting area, and any probable combination of primary loads resulting in the highest stresses in the supporting member should be used as design loads. Proper provisions shall be made for dynamic loading caused by wind and earthquakes. In any case, the loads other than dead load shall be not less than those recommended in the American National Standards Building Code Requirements for Minimum Design Loads (ANSI-A58.1). After all primary loads are defined, then, a series of possible design load combinations can be arrived at for inclusion in the structural analysis.

3. Preliminary Design

During this design stage, the design objective is to arrive as closely as possible at an optimum design that maximizes the serviceability and reliability, and at the same time minimizes the construction cost as well as service and maintenance costs.

With this goal in mind, a preliminary design strategy must be developed by an experienced structural engineer based on past experience and statistical data, which lead the design process toward an optimum solution.

In general, there are two extreme starting design strategies: (1) simple beam strategy (upper bound), and (2) fixed-beam strategy (lower bound). For the simple beam strategy, one assumes that all beams are simply supported, and the preliminary design can be started by selecting beam and column member sizes based on their simple moments and axial loads, respectively. During this iteration process, which will be introduced later, each subsequent member selection reduces the beam size and increases the column size as long as the design stresses of the beam are less than the code allowables. On the other hand, one could use fixed-beam strategy and assume that all beams are fixed-ended, except those that are actually hinged. Preliminary design, then, can be started by selecting beams based on the fixed end moments alone, and columns based on the axial loads combined with the fixed-end moments. In the iteration process, each subsequent member selection increases the beam size and reduces the column size as long as the design stresses in the columns are less than the code allowables. The preliminary member sizes as selected are all stored in the computer member data bank for future update during the "code-check" process.

To facilitate cost-effective construction procedures, the designer should group all structural members according to type (i.e., beams, columns, and tension members) and select one member size per group by the most critical design load combination within the group. For better utilization of the structural rolled shapes, one should establish computer files for beam member properties, column member properties, and tension member properties. The first two files are available from the AISCM.

4. Analysis

A complete structural analysis should be performed by computer as soon as the geometric configuration, design' loadings, boundary conditions, and preliminary member shapes and sizes are defined. This analysis can be performed on IBM, CDC, and other computer systems using well-known computer programs for structural analysis, such as STRUDL-DYNAL, NASTRAN, ANSYS, NISA, STARDYNE, SAP, SPACE, EASE, and STRESS. These computer programs can be accessed through large computer service vendors such as McDonnell Douglas Automation (MCAUTO), Control Data Cybernet Services (CDC), and Boeing Computer Services (BCS). Some of their programs can be purchased for a fee. SPACE is solely licensed through its developer, Digital Analysis Consultants. Some of the computer programs mentioned above can do both static and dynamic analyses, with varying degrees of nonlinear capabilities. All except STRESS can accept both prismatic members and finite elements.

5. Code Check

As previously introduced in process 3, preliminary design, there are two starting design strategies for initial member-size selection. The code-check conditional iteration is illustrated in Flowchart 11.5 and is required to ensure that members meet the code requirements. For the correction/optimization

Flowchart 11.5

Code-Check Conditional Iteration

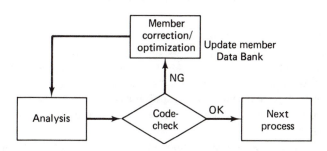

of member selection during the iteration process, each subsequent member selection chooses smaller beams with larger columns for the simple beam strategy, or larger beams with smaller columns for the fixed-beam strategy. This iteration process revises the member group data bank until every member group contains the minimum member size that meets the code requirements.

6. Connection Design

As soon as all structural members are defined, the connection detail design can begin. For simplicity, economy, and uniformity in fabrication and erection of connections in steel construction, they should be separated into two groups: (1) shop connections, and (2) field connections. The typical connections in each group can then be designed on the basis of the critical conditions involved within the group by the aid of the AISC Manual, or by a special computer subroutine.

7. Issue Construction Specifications and Design Drawings

Construction specifications and design drawings constitute the final step in the design process. The construction specifications, in general, include all material strengths and grades, construction methods and procedures, site-testing requirements, and inspection requirements. Design drawings are the final design products, including foundation, floor and roof plans, elevation views, cross sections, connections, and special details.

Construction specifications can be generated by a computer by editing a complete general specification which previously has been stored in a computer file. The editing consists of changing, adding, and deleting from the file. High-quality design drawings (ink plot) can be produced rapidly by an autographic computer system, consisting of a minicomputer, digitizers, CRT (cathode ray tube) terminal, input device, and output plotter.

11.4 COMPUTER-AIDED OPTIMIZATION

The design objective is to simultaneously maximize structural appearance, serviceability, and reliability, and at the same time minimize total cost of design, construction, and maintenance. In the computer-aided technology, one should be able to vary the primary design parameters, such as column and floor spacings, materials, and structural systems, to arrive at an optimum design in terms of merit (worth) and cost (expenditure), as illustrated in Fig. 11.3.

An experienced structural engineer with statistical data can simplify a complex structure to a manageable number of prime design parameters for rapid analytical processing. In order to arrive at an optimized overall solution, as illustrated in Fig. 11.4, the overall design optimization generally requires simultaneous consideration of the following primary factors: (1) materials, (2) configuration, (3) loadings, and (4) codes.

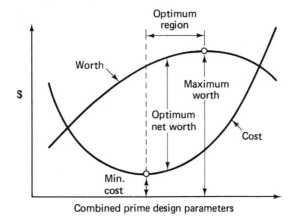

Fig. 11.3 Optimum net worth.

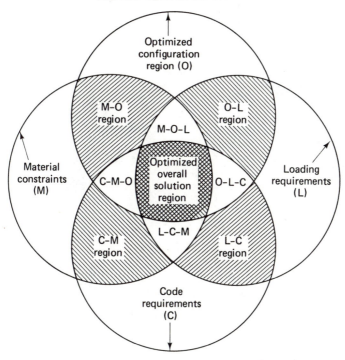

Fig. 11.4 The concept of overall design optimization.

INDEX

Date D